D1622262

EARTH'S FURY

An Introduction to Natural Hazards and Disasters

ROBERT L. KOVACH
Stanford University

No Longer
the Property of
Bluffton University

Prentice Hall, Englewood Cliffs, New Jersey 07632

Library of Congress Cataloging-in-Publication Data

KOVACH, ROBERT L. (ROBERT LOUIS)
    Earth's fury: an introduction to natural hazards and disasters/
Robert L. Kovach.
        p.   cm.
    Includes bibliographical references and index.
    ISBN 0-13-042433-1
    1. Natural disasters.   I. Title.
GB5014.K68   1995   94-32127
363.3'4—dc20   CIP

Acquisitions editor: *Robert McConnin*
Editorial/production supervision and interior design:
    Kathleen M. Lafferty/Roaring Mountain Editorial Services
Proofreaders: *Bruce D. Colegrove, Maria McColligan*
Manufacturing buyer: *Trudy Pisciotti*
Cover designer: *Bruce Kenselaar*
Front cover: *Spencer Grant/Photo Researchers, Inc.*
Back cover: *Bob Strong/Sipa Press;*
    G. Brad Lewis/Liaison International

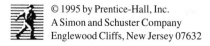 © 1995 by Prentice-Hall, Inc.
A Simon and Schuster Company
Englewood Cliffs, New Jersey 07632

All rights reserved. No part of this book
may be reproduced, in any form or by any means,
without permission in writing from the publisher.

Printed in the United States of America

10   9   8   7   6   5   4   3   2   1

ISBN 0-13-042433-1

Prentice-Hall International (UK) Limited, *London*
Prentice-Hall of Australia Pty. Limited, *Sydney*
Prentice-Hall Canada Inc., *Toronto*
Prentice-Hall Hispanoamericana, S.A., *Mexico*
Prentice-Hall of India Private Limited, *New Delhi*
Prentice-Hall of Japan, Inc., *Tokyo*
Simon & Schuster Asia Pte. Ltd., *Singapore*
Editora Prentice-Hall do Brasil, Ltda., *Rio de Janeiro*

To Linda, Denise, Dianne, John, and Robert

Contents

7 Atmospheric Hazards   100

8 Oceanographic Hazards   117

9 River Floods   133

10 Some Accident Scenarios   146

Preface

For many years I taught a course on earthquakes at Stanford University. That course was eventually enlarged to cover many other natural hazards in a way that is reasonably accessible to a broad spectrum of students. In preparing my class notes over the years, I have had to use parts of many different technical books because there is no clear and concise text on natural hazards and their risks. This book was written to fill that gap.

Earth's Fury: An Introduction to Natural Hazards and Disasters is designed to meet the needs of students who enroll in a basic course on natural hazards, human affairs, risk assessment, prevention, and mitigation. It is not surprising that courses dealing with natural hazards and their relation to earth systems science are popular. This book is a current and comprehensive exposition, written for an introductory audience but also useful as an advanced technical review. Some rather technical and mathematical information is inevitable in any book that deals seriously with this subject, but very little previous mathematical knowledge is necessary to grasp the demonstrations of probabilistic methods of risk assessment in the chapters. Each chapter concludes with review questions and problems that highlight major concepts; answers are provided at the end of the book. Appendices contain explanations of some basic mathematical concepts and a technical analysis of volcanic and seismic risk.

Not all natural hazards are discussed here. The choices reflect my personal interests somewhat. For instance, a good deal of material about earthquakes is included, not only because the possibilities of prediction and control in this field are exciting, but also because I have experienced earthquakes first-hand. In July 1952 I was sleeping on a bed with casters when the magnitude 7.7 Kern County earthquake struck in the early hours of the morning. My bed smashed into the wall, and my fate as a future geophysicist was sealed. In October 1989 I was in my study when the magnitude 7.1 Loma Prieta earthquake violently shook my house, causing me to cower beneath my desk.

On sabbatical in Hawaii, while a visitor at the Hawaii Institute of Geophysics, I witnessed some spectacular sunsets and sharpened my thinking about volcanic eruptions. Much of my professional life has been spent doing geophysical surveys in the deserts of the U.S. Southwest, Mexico, Pakistan, and the Middle East, where I observed the phenomenon of desertification and had a curious encounter with a dust devil.

I have been struck by the large number of natural disasters in recent decades. It is hard to tell whether this is a function of population growth, a cyclical phenomenon, or the result of an increase in natural hazards. Greater efforts at analyzing and mitigating earthquakes, tsunami, landslides, tropical cyclones, volcanic eruptions, and other disaster-making hazards have been under way since the United Nations General Assembly declared the 1990s as the International Decade for Natural Disaster Reduction. There is a new sense of urgency about identifying hazard sites and estimating the risk of disaster at those sites. The violent eruptions of two volcanoes on opposite sides of Rabaul harbor in Papua New Guinea during late September 1994, which prompted the evacuation of 30,000 residents, has underscored the continued vigilance of potential natural hazard sites. I hope that readers of this book will also develop a sense of urgency about natural hazards and what we can do about them.

Several people helped me in my search for reference materials and photographs. In particular, I thank Charlotte Derksen and John Baltierra of the Branner Earth Sciences Library at Stanford for the countless times they opened the locked stacks for me and pointed me in the right direction, and Michael Moore of the U.S. Geological Survey for giving me a wealth of historic photographic material. Gregory C. Beroza suggested that I integrate my class notes into a readable text. James E. Slosson gave valuable insight for Chapter 10 on risk analysis. Thanks to Kathleen Lafferty of Roaring Mountain Editorial Services for her help in configuring a finished product. Final thanks go to my wife, Linda, for her computer expertise and constant encouragement.

Robert L. Kovach

1

Introduction

This book is about *natural* hazards—those that require the right conditions in the earth's atmosphere or on or under its surface to become **disasters**, calamitous events that cause enormous destruction of life and/or property. A disaster can be sudden or slow. Earthquakes and volcanoes, floods and landslides, discharge their massive amounts of energy suddenly and inflict vast destruction very quickly. Droughts and desertification destroy more slowly and inexorably.

Natural hazards are often amplified by human actions. The population explosion has swelled the numbers of people moving into floodplains, earthquake-prone regions, coastal areas vulnerable to tropical cyclones, storm surges, and tsunami, and semiarid lands susceptible to drought and land degradation. Although our advanced technology allows us to protect ourselves from the vagaries of nature to an extent that would make our ancestors marvel, paradoxically our vulnerability to natural disasters may be increasing.

Human life has always depended on energy, but whereas our ancestors merely lit a fire for warmth and cooking, our standard of living demands gas and oil pipelines and hydroelectric, coal-fired, and nuclear-powered energy plants—all of which present new hazards. We build dams and levees to offset the cyclical effects of floods, but as the 1993 deluge in the Midwest showed, our elaborate flood-control systems may increase the magnitude of a flood by obstructing and channeling the wild river waters too rigidly. Our complex civilization makes us vulnerable to natural disasters in other ways. In simpler times an earthquake, volcano, or flood might have disrupted transportation and energy systems for a short time, but because these systems were rather primitive, they were easy to reconstruct. Today damage to bridges, roads, and tunnels may take years, and enormous sums, to repair. And the wreckage of a gas pipeline, a waste disposal facility, or a nuclear power plant can cause great havoc.

Still, few of us would want to exchange our vulnerabilities for those of our ancestors. We have a far more secure life in most ways. Science has given us a much fuller understanding of natural hazards so that we are growing more and more adept at identifying and evaluating them, and thus at making objective decisions about the risks they pose and how to mitigate them.

1-1 HAZARDS AND DISASTERS

Table 1-1 lists some of the worst natural disasters of the twentieth century and the resulting fatalities. Using this list, which is representative of the types of natural hazards that are considered rapid-onset disasters, we can state some worldwide fatality statistics on an annualized basis.

From 1900 to 1993, there were about 14,000 annual fatalities from earthquakes and 800 from volcanic eruptions and their consequences. Tsunami—seismic sea waves initiated by earthquake or volcanic eruptions—also caused many fatalities, particularly around the Pacific rim. About 9,000 fatalities a year resulted from tropical cyclones, typhoons, and hurricanes. (These names all refer to the same severe storm phenomenon. *Tropical cyclone* is the general term; *hurricane* is used in the western Atlantic, and *typhoon* in the western Pacific. In Australia a tropical cyclone is called a "willy-nilly," in the Philippines a "baguio," and in Arabic lands an "asifat.")

The Bay of Bengal coasts of India and Bangladesh are repeatedly battered by tropical cyclones; this is, in fact, the most dangerous region in the world for this natural hazard. About 15% of the world's tropical cyclones start either in the Bay of Bengal or the Arabian Sea. Of all the nations in the world, the Philippines has experienced the most natural hazards in the twentieth century. India was a distant second, followed by the United States in third position.

Note that Table 1-1 lists only sudden onset natural disasters. Other natural hazards of longer duration, such as drought and desertification, ultimately lead to famine, sickness, and death. In sub-Saharan Africa the death toll from these gradual destroyers has been staggering. Note also that the death tolls from such human-created hazards as nuclear and industrial plant accidents can be severe.

1-2 SOME DEFINITIONS

At this point it is useful to introduce a few definitions. What exactly do we mean by "hazard"? We define a **hazard** as a source of danger whose evaluation encompasses three elements: the risk of human harm, such as injury, trauma or death; the risk of property damage; and the acceptability of the level or degree of risk. Evaluating a potential natural hazard involves a sequence of considerations. A phenomenon must

Table 1-1   Some Sudden-Onset Natural Disasters of the Twentieth Century

Year	Event	Location	Approximate death toll
1900	Hurricane	United States	6,000
1902	Volcanic eruption	Martinique	29,000
1902	Volcanic eruption	Guatemala	6,000
1906	Typhoon	Hong Kong	10,000
1906	Earthquake	Taiwan	6,000
1906	Earthquake/fire	United States	1,500
1908	Earthquake	Italy	75,000
1911	Volcanic eruption	Philippines	1,300
1911	Floods	China	100,000
1915	Earthquake	Italy	30,000
1916	Landslide	Italy, Austria	10,000
1919	Volcanic eruption	Indonesia	5,200
1920	Earthquake/landslide	China	200,000
1922	Tropical cyclone	China	50,000
1923	Earthquake/fire	Japan	143,000
1926	Hurricane	United States	372
1928	Hurricane/flood	United States	6,000
1930	Volcanic eruption	Indonesia	1,400
1932	Earthquake	China	70,000
1933	Tsunami	Japan	3,000
1935	Earthquake	India	60,000
1935	Hurricane	United States	400
1938	Hurricane	United States	600
1939	Earthquake/tsunami	Chile	30,000
1944	Hurricane	United States	389
1945	Floods/landslides	Japan	1,200
1946	Tsunami	Japan	1,400
1948	Earthquake	USSR	100,000
1949	Floods	China	57,000
1949	Earthquake/landslide	USSR	12,000–20,000
1951	Volcanic eruption	Papua New Guinea	2,900
1953	Floods	North Sea coast (Europe)	1,800
1954	Landslide	Austria	200
1954	Floods	China	40,000
1954	Hurricane	Haiti and United States	347
1955	Hurricane	United States	400
1957	Hurricane	United States	526
1959	Typhoon	Japan	4,600
1960	Earthquake	Morocco	12,000
1961	Typhoon	Hong Kong	400
1962	Landslide	Peru	4,000–5,000
1962	Earthquake	Iran	12,000
1963	Tropical cyclone	Bangladesh	22,000
1963	Volcanic eruption	Indonesia	1,200
1963	Landslide	Italy	2,000
1964	Floods	Vietnam	8,000
1965	Tropical cyclone	Bangladesh	17,000

Table 1-1   (Continued)

Year	Event	Location	Approximate death toll
1965	Tropical cyclone	Bangladesh	30,000
1965	Tropical cyclone	Bangladesh	10,000
1968	Earthquake	Iran	12,000
1970	Earthquake/landslide	Peru	70,000
1970	Tropical cyclone	Bangladesh	300,000–500,000
1971	Tropical cyclone	India	10,000–25,000
1972	Earthquake	Iran	5,400
1972	Earthquake	Nicaragua	5,000
1974	Hurricane	Honduras	8,000
1975	Earthquake	Turkey	2,400
1976	Earthquake	China	250,000
1976	Earthquake	Guatemala	24,000
1976	Earthquake	Italy	900
1976	Earthquake	Philippines	3,600
1976	Earthquake	Turkey	3,600
1977	Tropical cyclone	India	20,000
1978	Earthquake	Iran	25,000
1979	Hurricane	Caribbean	1,400
1980	Earthquake	Algeria	2,600
1980	Earthquake	Italy	3,100
1982	Earthquake	Yemen	3,000
1982	Volcanic eruption	Mexico	1,700
1983	Earthquake	Turkey	1,300
1984	Typhoons	Philippines	1,000
1985	Tropical cyclone	Bangladesh	11,000
1985	Earthquake	Mexico	10,000
1985	Volcanic eruption	Colombia	23,000
1986	Volcanic gas eruption	Cameroon	1,700
1986	Earthquake	El Salvador	1,000
1987	Earthquake	Ecuador	1,000
1987	Floods	Bangladesh	1,600
1987	Wildfire	China	200
1987	Floods	Nepal	400
1988	Floods	Bangladesh	3,000
1988	Earthquake	Armenia	50,000
1989	Typhoon	Thailand, India	1,000
1989	Earthquake	United States	61
1990	Earthquake	Iran	50,000
1991	Tropical cyclone	Bangladesh	138,000
1991	Volcanic eruption	Philippines	1,000
1992	Hurricane	United States	20
1992	Hurricane	Kauai	3
1992	Floods	Pakistan	2,000
1992	Earthquake	Egypt	500
1992	Landslide	Bolivia	210
1992	Earthquake	Indonesia	1,200
1993	Tropical cyclone	Fiji	25

Table 1-1   (Continued)

Year	Event	Location	Approximate death toll
1993	Avalanche	Turkey	300
1993	Volcanic eruption	Philippines	100
1993	Earthquake	India	9,700
1994	Earthquake	United States	50

SOURCE: *Confronting Natural Disasters* (National Academy Press, Washington, D.C., 1987). Updated by author.

first be *perceived* and then *identified* as a hazard. The next step is *risk evaluation*. The sequence ends with the *assessment* and *possible control* of the identified hazard's potential outcomes.

A key ingredient in the assessment and control of hazards is the concept of a **safety threshold**—the point above which a phenomenon constitutes a hazard. Setting a threshold is equivalent to specifying the level of risk that is socially tolerable. If the *catastrophic* threshold of risk is set too high, the ability to detect lower-level risks will be compromised. Data acquisition is costly, and at times information is deliberately mishandled. The extreme outcome of threshold manipulation is to find no risk at all from a hazard! The great majority of deaths from natural disasters occur in poor and developing countries, where risk assessment, safety thresholds, and mitigative measures are least likely and where governments are least able to respond to disasters.

Because we cannot avoid or prevent all natural hazards, our goal must be the minimization or mitigation of their effects. **Mitigation** has physical, engineering, and social aspects. The physical aspects involve collecting and analyzing data and using this information to develop theories that will lead to better predictions of the consequences of hazards. Although there will always be an element of uncertainty in our assessments, this kind of technical knowledge is essential for implementing the second aspect of mitigation: engineering countermeasures such as the strengthening of bridges and buildings, the construction of safe dams, and the erection of avalanche barriers. Finally, the social aspects of mitigation encompass evaluation of risks and the economic costs of reducing or eliminating those risks.

1-3 EARTH'S FURY

Probably half of the world's people live in areas that are highly vulnerable to natural disasters. We in the United States have been through two catastrophic hurricanes, a major earthquake, a flood unprecedented in our history, and devastating wind-generated fires since 1990.

Life means risk. Periodically we are going to be overmatched by earth's fury: the engulfing storm surge, the wild river floods, the cataclysmic earthquake, the shattering landslide, the awesome volcano. We cannot avoid all natural hazards, but we can seek to understand them so we can better deal with them.

The chapters that follow take up the primary severe natural hazards one by one and sketch what is known about their causes. Their sometimes disastrous effects are related in vivid examples taken from history, and mitigative measures are outlined. Earthquakes are covered more extensively than other natural hazards, both because they produce a greater number of fatalities in an average year than any other natural hazard, and because they are an outstanding example for our purposes. We know a lot about their genesis and development, and we have a good deal of experience with their effects in this country, especially in California. Mitigative measures are well-tried and successful. Moreover, current work on earthquake prediction is provocative and promising.

We begin, however, with volcanoes.

REVIEW

1. What is a hazard?
2. List the considerations necessary for evaluating a potential natural hazard.
3. Why is the setting of a safety threshold for a hazard important? What is the chief danger in setting a catastrophic threshold of risk?
4. Which two sudden-onset natural hazards produce, on an annualized basis, the greatest fatality rates?

2

Volcanoes

As if to celebrate our arrival the volcano saluted us by an eruption. We saw the abyss kindling at our feet, whilst a magnificent jet of fire rose towards us with a noise resembling the rapid discharge of artillery. The night had long since closed around us, our guides were urgent that we should make the descent: we were therefore compelled to yield to their request and prepare for our return; but before we made a final retreat we waited to see another eruption, and this, fortunately, proved to be most magnificent. The three mouths were playing simultaneously, and reflecting the reddish brightness of the lava; whilst the triple enclosure of the crater revealed itself once more to our eyes.

Like this nineteenth-century observer of Stromboli, most of us are fascinated by volcanoes. These spectacular phenomena have produced some of the most beautiful landforms on earth, such as the Caribbean Islands, Hawaii, and the Cascade Range of the Pacific Northwest. Yet there is a risk to living on or near a geologic feature shaped by volcanic eruptions, for the furious outbursts may be repeated. Nearly 10% of the world's people live close to volcanoes. Although human fatalities from volcanic eruptions are much lower than those from earthquakes and tropical cyclones, the economic consequences can be severe.

2-1 UNDER THE VOLCANO

The most famous volcanic eruption of ancient times was the explosion of Italy's Mount Vesuvius in A.D. 79 that buried the Roman towns of Pompeii and Herculaneum under a barrage of mud and hot ash. About 16,000 people died in Pompeii. Some of the bodies and portions of the buildings were preserved, forgotten, in the dry ash until rediscovered in the late sixteenth century.

In 1815 Mount Tambora on Sumbawa Island, Indonesia, exploded, killing 12,000 people and throwing large quantities of ash and dust into the atmosphere, which disrupted global weather patterns and devastated agriculture in Europe and Canada.

In 1883 Krakatoa (Krakatu) near Java, Indonesia, produced the most devastating volcanic eruption in recorded history. Enormous water waves generated by the undersea explosion rolled over nearby islands, including Java and Sumatra, drowning

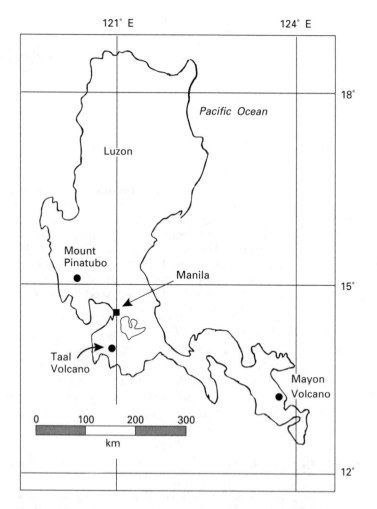

Figure 2-1   Map of the island of Luzon, Philippines, showing the locations of the Mount Pinatubo, Taal, and Mayon volcanoes.

36,000 people. The dust cloud from Krakatoa covered the earth for years, intercepting incoming sunlight and noticeably lowering global temperatures.

In 1980 Mount St. Helens, situated in the Cascades in the State of Washington, blew up, producing one of the worst natural disasters this country has ever seen. The volcano had been dormant since 1857, and local residents had learned to live with it casually. The great blast destroyed beautiful age-old forests and loosed giant mud-flows that raced downslope, creating floods and wiping out bridges, homes, and industrial complexes. Estimates of property damage were as high as $2 billion.

From April 2 to June 18, 1991, there was a series of volcanic eruptions from Mount Pinatubo in the Philippines (Figure 2-1). The casualty total will probably nev-er be known, but there were at least 1000 dead or injured. Many of the casualties were caused by collapsing roofs laden with volcanic ash soaked by Typhoon Yunya, which passed a short distance north of Mount Pinatubo on June 15. The typhoon was a sec-ond natural disaster, extending the deposition of ashfall to many areas of the islands that did not expect it, such as the capital city of Manila. Clark Air Base, a U.S. facility some 20 km east of Mount Pinatubo was closed down permanently because of the destructive ashfall. This volcanic disaster would have produced far more casualties had not 58,000 people been evacuated before the disastrous eruption on June 15 because a rational, though uncertain, forecast was given, believed, and responded to by local authorities.

We can draw several lessons from the Mount Pinatubo experience. First, increased unrest at a volcano thought to be dormant for centuries must be heeded. Second, uncertain forecasts, though they may prove politically embarrassing are valu-able for assessing preparedness and warning people. Forecasts of eruptions are sel-dom certain, but trained observers using seismic, temperature, and other scientific measures have achieved a fair degree of accuracy—and have saved many lives by giv-ing early warnings. Third, time-sequential upgraded levels of volcano monitoring are needed. Finally, past eruption histories give definite clues to the future behavior of a volcano.

2-2 TYPES OF VOLCANOES

A **volcano** is an opening or vent in the earth's crust (outer layer) from which an under-ground reservoir of magma rises to the surface and erupts or flows out as lava. **Magma**, rock located deep within the crust that contains considerable quantities of dissolved gases, is under such intense heat and pressure that it is plastically deformed and flows like a liquid. When a chamber of this molten rock connected to a surface vent mixes with underground water, the water is changed into a gaseous state that may expand until the pressure forces it through the vent. The resulting volcanic eruption, called **lava**, is a mixture of the magma and the condensed steam produced under-ground. Note that although we use the word "eruption" to describe this release of magma, it sometimes takes the form of an ooze rather than an explosion.

Volcanoes may be classified in several ways. One is by their activity status.

Active or *live volcanoes* are those that erupt regularly (every few years) or even continually. *Dormant volcanoes* erupt less frequently—perhaps every half century or even less. *Extinct volcanoes* are presumed to be dead—that is, incapable of erupting again, however violent their past. In practice, it is often hard, and sometimes dangerous, to distinguish between dormant and extinct volcanoes.

Another classification system uses the type of surface opening as the criterion: *Central eruptions* are those from a central vent, *fissure eruptions* issue from cracks or fissures, and *areal eruptions* are scattered across a wide surface. It is also possible to classify volcanoes by the type of lava they produce or by the violence of their eruptions.

We are chiefly concerned with the types of hazards volcanoes present. Let us now look at these.

2-3 TYPES OF VOLCANIC HAZARDS

Hazardous volcanic events will be classified as (1) pyroclastic falls, (2) pyroclastic flows and debris avalanches, (3) lahars, (4) lava flows, and (5) volcanic gases.

Technically, **pyroclastics** refers to fragmentary *rocks* ejected from a volcanic vent, often at great volume, while **tephra** refers to *all* types of material expelled by a volcano, including pyroclastics. In practice, the terms are often used interchangeably, and they will be so used here. *Pyroclastic* or *tephra falls* are a combination of rock fragments and partially or wholly solidified lava that is spewed into the air by a volcanic explosion. The size of the ejecta further describes the phenomenon. If the particles are less than 2 mm in diameter, they are referred to as **ash**. Fragments ranging from 2 to 64 mm are called **lapilli**. Volcanic **bombs** are larger than 64 mm—the size of an orange. Typically, bombs are the size of a baseball or volley ball, but some are so enormous that they weigh many tons. This size classification is not meant to be scientifically rigorous but rather usefully descriptive. Figure 2-2 shows the location of La Soufrière, a volcano on the Caribbean Island of Guadeloupe that erupted in 1976. The eruption, captured in the photograph in Figure 2-3, consisted of volcanic vapors and volcanic ash.

The hazards from a pyroclastic fall are substantial. These falls can bury structures, ruin agriculture, reduce visibility over large areas, and cause secondary fires. The misery can be compounded by atmospheric and climatic effects, such as wind and rainfall that carry the pyroclasts great distances. Pyroclastic falls have produced about 4% of all volcano-related deaths since 1600.

Pyroclastic flows are the most disastrous form of volcanic eruption. In these explosions there is a tremendous release of pressure resulting in the violent expulsion of a superheated mixture of gas and rock. The directed blast can produce a pyroclastic surge of **nuée ardente** ("glowing cloud" or "glowing avalanche"), a gravity-controlled mixture of gas and solid materials that can attain temperatures of 350° to 1000°C and has a velocity ranging from 100 to 600 km/hr. This scalding avalanche destroys flesh, buildings, landscapes, and anything else in its path through its impact, heat, abrasion, and toxic gases. *Debris avalanches* are produced by the large-scale

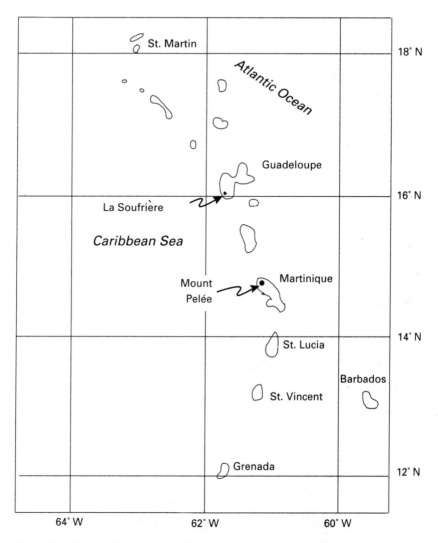

Figure 2-2   Map of the Lesser Antilles showing the location of the volcanoes La Soufrière on Guadeloupe and Mount Pelée on Martinique.

collapse of a portion of a volcano. They often precede a directed blast from the volcano's side.

One of the most devastating nuées ardentes was the explosion of Mount Pélee on the West Indian island of Martinique (Figure 2-2) in 1902. The temperature of the fiery cloud that blasted a mixture of dust, steam, and gas into the town of Saint-Pierre is estimated to have been as high as 1000°C. Within minutes the town was a smoldering ruin and 30,000 people were dead.

Figure 2-3 La Soufrière is a volcano on the island of Guadeloupe in the Caribbean that has explosively erupted about 10 times since 1400. It last erupted in 1976, prompting the evacuation of 70,000 people. The view shows the plume of vapor and ash. (Photograph courtesy F. C. Whitmore, U.S. Geological Survey.)

Pyroclastic flows have been responsible for 21% of all volcano-related deaths since 1600. Figure 2-4 shows an eruption of Mayon Volcano on the island of Luzon in the Philippines. Mayon has erupted explosively about 40 times since 1616, producing many nuées ardentes, and is the most active volcano in the Philippines. When it exploded in 1993 after an 8-year quiescence, volcanic ash was thrown 4 to 5 km into the sky. A portion of the volcano sheared off, disgorging tons of superheated material on farming communities.

Lahars are mudflows containing rock debris and blocks of predominantly volcanic origin mixed with water. Essentially, lahars are flows of pyroclastics mobilized by water. These rapid mass flows originate on the slope of a volcano when hot volcanic ejecta mixes with the ice or snow covering the slope (Mount St. Helens) or when ash flow deposits are turned into mud by heavy rains (Mount Pinatubo). Lahars are controlled by the topography, which directs and restricts their flow. Downslope velocities can range from 25 to 200 km/hr. About 14% of all volcano-related deaths since 1600 have been caused by lahars.

Figure 2-4   Explosive eruption of Mayon Volcano in the central Philippines on May 2, 1968. The photo shows the vertical ejection of incandescent material to a height of 600 m and the downfall feeding nuée ardente. (Photograph courtesy J. G. Moore, U.S. Geological Survey.)

More than 22,000 people were killed when the snow-covered volcano of Nevado del Ruiz in the Andes Mountains in Columbia erupted on November 13, 1985 (Figure 2-5). The hot volcanic ash melted part of the icecap, triggering lahars that flowed down river valleys and overran villages in their paths. A devastating lahar covered the city of Armero, where 84% of the 25,000 residents perished. This lahar produced history's fourth-largest death toll from a single eruption.

Kilauea Volcano on the island of Hawaii (Figure 2-6) is one of the best-studied volcanoes in the world. Its spectacular eruptions produce two types of *lava flows*: **pahoehoe**, which has a smooth, billowy, or ropy surface; and **aa**, which has a rough, jagged, spinous, clinkery surface that may contain large polyhedral blocks (Figure 2-7).

Lava flows are slow moving, controlled by topography. The typical flow is 1 to 10 m thick and travels tens of kilometers. Because their slow velocity gives people plenty of warning of their approach, lava flows have been responsible for less than 1% of volcano-related deaths, though they can do great damage to land and property (Figure 2-8). On the island of Hawaii most people consider lava flows an acceptable

Figure 2-5 A steam eruption of Nevado del Ruiz, Colombia, in September 1985 prior to the major eruption on November 13, 1985. (Photograph courtesy U.S. Geological Survey.)

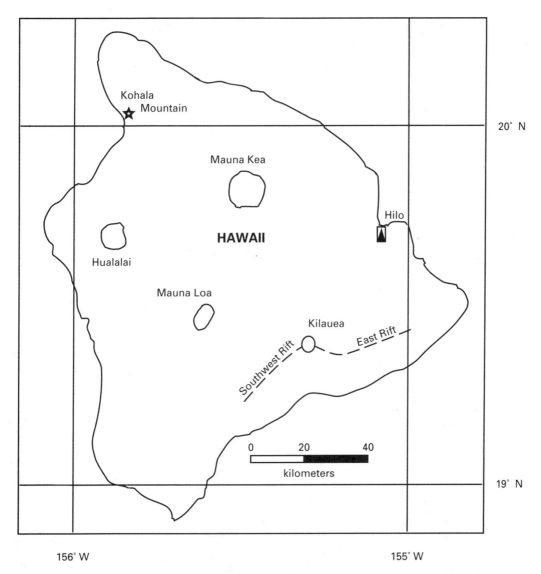

20° N

Kohala
Mountain

Mauna Kea

Hilo

HAWAII

Hualalai

Mauna Loa

Kilauea

East Rift

Southwest Rift

0   20   40

kilometers

19° N

156° W   155° W

Figure 2-6   Map of the island of Hawaii showing Kilauea and Mauna Loa volcanoes.

risk because they do not cause many fatalities and the affected land is usually safe for
human habitation after a few decades.

The last type of volcanic hazard we shall consider is *volcanic gases*. Volcanoes
can eject a variety of toxic gases. In August 1986 carbon dioxide (CO_2) and other
gases were emitted from the bottom sediments of Lake Nyos in Cameroon, Africa.
Lake Nyos was long believed to be a dormant volcanic crater filled with water, but it

Figure 2-7 Aa flow moving over earlier pahoehoe flow on the big island of Hawaii on February 15, 1990. Observe the clinkery upper surface of the aa flow and the ropy surface of the pahoehoe flow. (Photograph courtesy J. D. Griggs, Hawaii Volcano Observatory, U.S. Geological Survey.)

Figure 2-8 Road closed—one consequence of a recent lava flow on the island of Hawaii. (Photograph by Robert L. Kovach.)

Table 2-1   Human Fatalities from Volcanic Activity, 1600–1986

Primary cause of fatalities	Number of fatalities	Percent of all volcano-related fatalities
Pyroclastic flows	55,000	21
Tsunami (volcano induced)[a]	44,000	16.8
Lahars	36,700	14
Tephra (pyroclastic falls)	11,000	4.2
Volcanic gases	1,900	0.7
Lava flows	1,000	0.4
Disease, starvation	95,300	36.3
Other or unknown	17,300	6.6
TOTAL	262,200	100

SOURCE: Condensed from Tilling (1989).
[a]Tsunami, which are discussed in Chapter 8, are large water waves in the ocean generated by earthquakes or major eruptions of undersea volcanoes or volcanic islands.

was apparently reactivated by a small eruption. Clouds of CO_2, which are heavier than air, hugged the low ground, causing death by suffocation to animals and about 1700 villagers. Other toxic gases emitted by volcanoes are sulfur dioxide (SO_2), hydrogen sulfide (H_2S), and carbon monoxide (CO). When SO_2, mixes with moisture in the atmosphere, it can form acid rain, which may damage crops over a widespread area. Volcanic gas emissions are responsible for about 0.7% of volcano-related deaths.

Volcanoes, which can exhibit their force in different forms, from violent explosions to inexorable slow flows, are a significant hazard to life and property. Table 2-1 is a tabulation of human fatalities from volcanic activity over the last four centuries. The highest percentages of fatalities have been produced by relatively infrequent explosions of volcanoes that were insufficiently monitored to give people enough warning, and by the disease and starvation brought on by the failure to respond effectively to the wreckage.

2-4 VOLCANO HAZARD ASSESSMENT AND MITIGATION

People have lived near active volcanoes from time immemorial for a number of reasons. Volcanic ash often produces fertile farmland, and volcanic rocks furnish excellent building materials. Economic benefits have been derived from mineral deposits associated with volcanoes, such as obsidian (a dark glass formed by the cooling of lava that was used by the ancients to make cutting tools), diamonds, and sulfides. Some volcanic areas have hot groundwater, a geothermal energy source for centuries whose use is increasing today. Living at the higher elevations of volcanoes provides relief from oppressive heat in the tropics and subtropics. Then, too, overcrowding pushes populations into volcanic regions. Finally, these regions often display great natural beauty. For all these motives—and because volcanoes can remain quiescent for long periods, lulling people into believing they are extinct or nearly so—many highly populated regions of Italy, Japan, Indonesia, the Philippines, and Central America are subject to volcanic hazards.

In assessing the potential hazards from a volcano and its possible mitigation, we need to consider the *volcano's status* (active, dormant, extinct), and *history*, its *frequency of eruption* or **recurrence interval** (the average time between repeated eruptions), and the concept of *volcano hazard zonation*. We discussed volcano status in an earlier section. These somewhat subjective and uncertain categories are of limited use in quantifying the probability and severity of a volcano hazard, but they do have documentary and descriptive value.

A far more important factor is recurrence interval. Certain volcanoes have a characteristic recurrence time. (In Appendix D we present a Poissonian probability model for calculating recurrence intervals of volcanoes.) Mount Etna in Italy and Kilauea in Hawaii, for instance, have recurrence intervals of tens of years. Other volcanoes have recurrence intervals of hundreds to thousands of years. Lake Nyos in Cameroon, which silently erupted with deadly carbon dioxide gas in 1986, has an unknown recurrence interval.

One of the most active volcanoes in the world is Stromboli, off the coast of Italy (Figure 2-9). The height of the volcano is about $\frac{1}{2}$ mile, with the principal active crater situated about two-thirds up from the base. Stromboli was described by many writers as an active volcano before the beginning of the Christian era, and it is in nearly continual eruption today. Most of these eruptions are minor (Figure 2-10), but occasionally Stromboli explodes with great force.

Pompeii was an ancient town at the southeastern foot of Mount Vesuvius in Italy (see Figure 2-9). In A.D. 63 an earthquake vented its force on the town. The inhabitants were still actively engaged in repairing the damage from that earthquake when, in A.D. 79, a truly cataclysmic eruption of Vesuvius covered Pompeii with small stones (lapilli), ashes, and cinders over the course of two days. Large amounts of steam given off by the volcano condensed with rain, which mixed with the volcanic ash and dust. A torrent of pasty mud (lahar) overwhelmed the town of Herculaneum at the western base of Mount Vesuvius. Unlike the inhabitants of Pompeii, however, people in Herculaneum had time to escape because the slower-moving lahar gave a generous warning compared with the eruption that obliterated Pompeii. After this first-century catastrophe, Vesuvius erupted intermittently until the year 1139. The volcano remained in a state of repose for nearly 500 years, until a violent explosion claimed about 18,000 lives in 1631.

Mount Vesuvius erupted spectacularly again in 1906, producing massive lava flows and an extensive deposition of volcanic ash a large distance from the main eruptive crater. Perhaps fearing a repetition of the "Last Days of Pompeii" of A.D. 79, 50,000 people fled. Still, there were about 200 fatalities and 6700 structures were damaged. Figure 2-11 shows the extent of the ashfall deposits and lava flows from the 1906 eruption. Some areas were permanently lost to agriculture. The area labeled II was covered by at least 50 cm of volcanic ash and was useless for crop production until

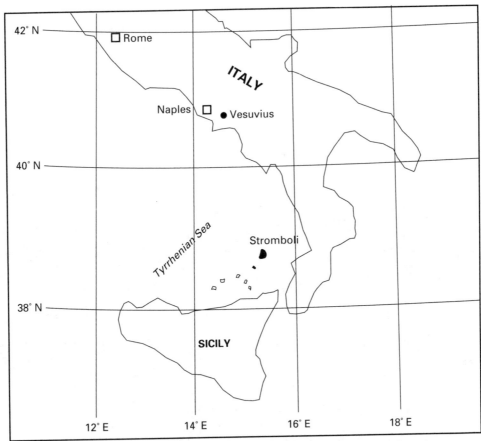

Figure 2-9 Map of the Italian region showing the locations of the volcano Stromboli in the Tyrrhenian Sea and Vesuvius near the city of Naples.

extensive tilling and fertilization improved soil conditions, which took at least 3 years. Areas farther away were less affected.

Can the effects of previous eruptions of a volcano give us insights into how it will behave in the future? In other words, does knowledge of a volcano's history give us a basis for zoning risk areas?

At the outset of this section we mentioned *volcano hazard zonation*. This is the concept that areas of potential hazard can be zoned according to the degree of risk they pose. Several factors are considered in making these estimates. One is the number of people exposed to danger, their locations, and their economic activities. Another key consideration is the location of infrastructure and emergency services and their susceptibility to hazard. A third factor is potential crop loss in cultures dependent on agriculture.

Figure 2-10   A night eruption of Stromboli, west of Italy, in March 1951. This volcano has had continuous eruptions for over 2000 years, some as frequently as every 15 to 30 minutes, giving it the nickname "Lighthouse of the Mediterranean." The phrase *Strombolian eruption* has been coined to describe the explosive discharge of incandescent lava. (Photograph by H. Williams, U.S. Geological Survey.)

Suppose we were to use Figure 2-11 as a basis for preparing a volcano hazard zonation map. We would need to bear in mind that there are now more than 500,000 people in the vicinity of Mount Vesuvius—quite a few more than in 1906. Estimates are that a replication of the 1906 eruption could produce ten times more fatalities. And this is not the worst case scenario. If the prevailing winds at the time of the eruption were directed toward Naples, a city of 1.2 million people, the fatality figure would be far higher. Volcano hazard zonation is still somewhat rudimentary, but at least it has the merit of focusing societies on preparing for a potentially severe event.

Some work has also been done on mitigating the effects of volcanic eruption. Countermeasures taken to protect against debris flows and lahars include the diversion of natural flow courses and the construction of dams to prevent or impede down-

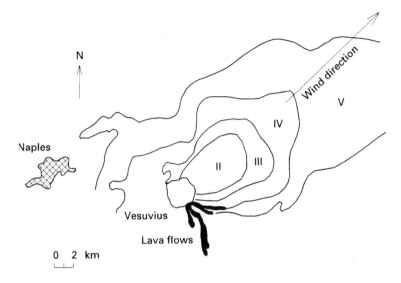

Figure 2-11   A map showing the distribution of lava flows and volcanic ash deposition from the 1906 eruption of Vesuvius. The area labeled II received ash deposits at least 50 cm thick; the area labeled III had ash deposits ranging in thickness from 15 to 50 cm. The higher numerals represent areas of lesser ash deposit.

stream movement. Drainage systems of pipes and tunnels have been put in place to lower the water level of reservoirs in the path of a flow, thus minimizing the danger of overtopping. There have been attempts to use pumped seawater and aerial bombing to cool slow-moving lava flows. Earth barriers have been constructed to redirect the downward flow, and fallout shelters have been built to shield populations from tephra deposits and ashfall. Finally, early warning systems, called *sabo works*, are already operational in Japan. When a sensor positioned high on the flank of a volcano detects the onset of a pyroclastic flow or lahar, it sets off a response in the sabo works that automatically stops road traffic, just as a signal at a railway crossing does.

REVIEW

1. What were lessons learned from the eruption of Mount Pinatubo in the Philippine Islands?
2. What is the difference between magma and lava?
3. Distinguish among active, dormant, and extinct volcanoes.

4. List and describe the major types of hazardous volcanic events.

5. What types of lava are found on the Island of Hawaii?

6. What is the most dangerous type of volcanic event in terms of human fatalities?

7. Why do people choose to live near volcanoes?

8. Name the four most important considerations for volcanic hazard assessment.

9. Describe the concept of volcanic hazard zonation.

3

Earthquakes

3-1 THE GEOGRAPHY OF EARTHQUAKES

When we look at a map of major earthquakes sites (Figure 3-1), we notice immediately that earthquakes are primarily confined to narrow bands rather than being randomly distributed around the globe. This is also true of volcanoes. There are exceptions, of course, but most of the world's earthquake and volcano activity (Figure 3-2) takes place in the circum-Pacific belt, known as the "Ring of Fire." Indeed, 80% of all earthquakes arise in this belt.

An **earthquake** is a sudden shaking or rupture in the earth caused by the release of accumulated stresses in the crust. The point within the earth where the rupture starts is known as the **focus** or *hypocenter*. Directly above it on the surface of the earth is a point called the **epicenter**. Earthquakes can be classified according to the depth of their foci. A **shallow-focus earthquake** has a focal depth of from 0 to 70 km. The focal depth of an **intermediate-focus** earthquake ranges from 70 to 300 km, and that of a **deep-focus earthquake** from 300 to 800 km. Focal depths greater than 800 km do not occur. Considering that the radius of the earth is 6371 km, this means that earthquake activity—or **seismic activity**—is confined to the outer 12% of the earth. Shallow-focus earthquakes account for 85% of all energy released by earthquakes. Intermediate-focus shocks contribute about 12%, and deep-focus shocks a mere 3%.

One of the most active sectors of the circum-Pacific belt is the zone stretching from the Komandorski Islands into Central Alaska. This zone is also characterized by

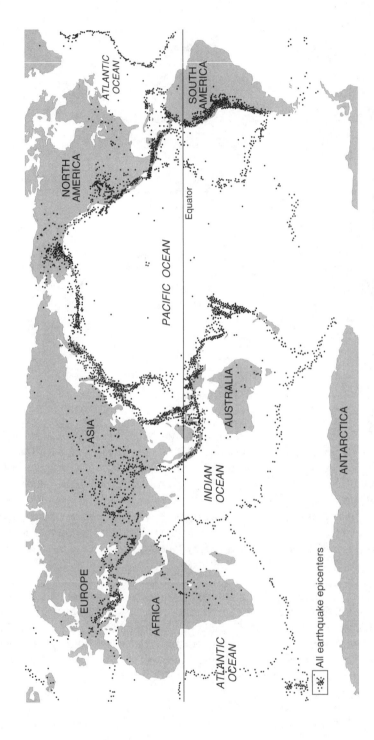

Figure 3-1 Global mosaic map of earthquake epicenters. (Courtesy of U.S. Geological Survey.)

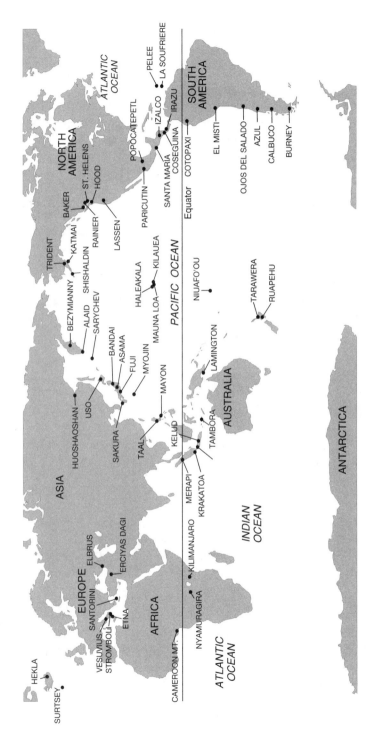

Figure 3-2 The location of some of the earth's prominent volcanoes. (Courtesy of U.S. Geological Survey.)

very high volcanic activity. As we follow the zone southeastward from Alaska to British Columbia, we find that seismic activity decreases to a moderate level, but farther south along the Pacific Coast it increases and continues to intensify into Mexico and Central America. Mexico and Central America have the greatest seismic activity in the western hemisphere as well as high volcanic activity. Shallow-focus earthquakes are frequent. Intermediate-focus earthquakes also occur in Central America, but there are no deep-focus earthquakes. From Yucatan (in southeastern Mexico) and Honduras the seismic zone traces an arc through the West Indies and loops around through Venezuela and Colombia. The entire Caribbean loop is seismically active, and the Lesser Antilles (part of the West Indies) are also prone to volcanic activity, such as was exhibited in the eruptions of Mount Pelée on the island of Martinique and La Soufrière on Guadeloupe.

The western part of South America is highly seismic, whereas the eastern part of the continent has a notable absence of seismicity. Shallow-focus shocks occur near the west coast, with a progression of intermediate- and deeper-focus shocks farther inland, beneath the Andes. This zone is typified by very large earthquakes, many of which have produced damaging tsunami and other large sea waves.

On the western side of the circum-Pacific belt are the Tonga and Kermadec zones, which are very active regions for shallow-focus earthquakes. Interestingly, fully half of the world's deep-focus shocks also take place here. Bathymetric measurements (measurements of water depth at various places in a body of water) have disclosed a deep-sea trench, parallel chains of islands, and the presence of volcanoes in the Tonga-Kermadec zone.

Seismic activity continues westward along the New Hebrides–Solomon Island chain to the equator off New Guinea. Here the seismic belt divides into two zones that progress northward to Japan—one along the Mariana Islands arc, and the other from the Philippines to Formosa. Japan and adjacent areas are part of the Kamchatka-Kurile Islands zone, the most active region in the world for shallow- and intermediate-focus shocks. Seismic events occur along the Indonesian arc to Burma and along the Himalayan belt north of India. (North of this belt into China, the pattern of seismicity becomes more diffuse.) The zone of seismic activity continues westward through Pakistan, Iran, and Turkey into the Mediterranean Sea region.

Earthquake-prone Greece, Albania, Romania, and Italy fall within an arc circling the Aegean Sea—an earthquake zone that accounts for about 15% of total global seismic energy. Other regions prominent in the global mosaic of earthquake geography are East Africa, the Gulf of Aden, the Red Sea, and the Dead Sea rift zone.

We conclude this discussion of earthquake geography with two observations about the global pattern of seismicity. First, seismic activity is greater in the northern hemisphere than in the southern hemisphere. Second, seismic activity decreases rapidly south of the equator—less than 10% of large earthquakes take place in the entire region below 30° south latitude, which constitutes about a quarter of the world's surface.

3-2 PLATE TECTONICS

Most of the world's earthquakes occur in well-defined belts that divide the earth's surface into large movable segments called **plates**. The boundaries of the plates, which
underlie the continents and oceans, are defined by these belts of seismic activity. The
plate's distribution does not correlate with the location of continents and oceans, however, because, in general, seismic belts do not follow the boundaries between continents and oceans. Differential movement takes place along the plate boundaries as a
result of large earthquakes that occur on the boundary. We call these **interplate
earthquakes** and, the seismic events that occur within the interior of the plates
intraplate earthquakes. *Plate tectonics* is the study of how these plates originated
and how they behave.

At this point we need to give a general idea of the structure of the earth. At the
center is the **core**, a fiery liquid mass thought to be of metallic iron surrounding an
inner solid mass thought to be of metallic nickel. Encircling the dense nucleus of the
core is an intermediate layer called the **mantle**. This solid layer ranges in depth from
tens of kilometers to 3470 km and consists of rocks of basic composition, such as
basalts and gabbros, that are rich in magnesium, iron, and calcium. The **crust** is the
outer layer of the earth. It has a thickness of tens of kilometers and is composed of
rocks rich in silica, such as granite and sandstone. Because the basic rocks of the mantle have a higher density than the silicic rocks of the crust, seismic waves have higher
velocities in the mantle than in the crust. The divisional boundary between crust and
mantle that marks this sudden change in seismic velocities is known as the
Mohorovicic or **M-discontinuity**.

What we call the **lithosphere** is the somewhat rigid shell of the earth consisting
of the crust and a portion of the upper mantle. This definition is based on rheological
rather than chemical considerations. (Rheology is the science dealing with the deformation and flow of matter.) The strong, rocky lithosphere is the material of which the
plates are made. Plate temperatures lie above the melting temperatures of rocks, and
the plates can endure considerable shearing stresses of hundreds of bars. Lithospheric
thickness beneath the ocean basins is about 70 km, and beneath the continents it
ranges from 100 to 200 km. All earthquakes occur in the lithosphere.

Underlying the lithosphere is the **asthenosphere**, a layer of material that has no
lasting endurance to shearing stress. The asthenosphere is very hot and partially molten.
It can sustain slow movement or **aseismic creep** along a fault without earthquakes.

The major plates—Pacific, Indo-Australian, Eurasian, North American, South
American, African, and Antarctic—were identified decades ago, but as the science of
plate tectonics has evolved, increasingly smaller plates have been delineated, such as
the Nazca, Caribbean, Philippine, Arabian, and Cocos plates. Figure 3-3 shows all the
plate tectonic boundaries thus far detected. It is the plate boundaries that are of most
interest to us because that is where the earthquake action is. There are four main
types: accretionary or divergent, subducting and collision (both of which are convergent), and transform.

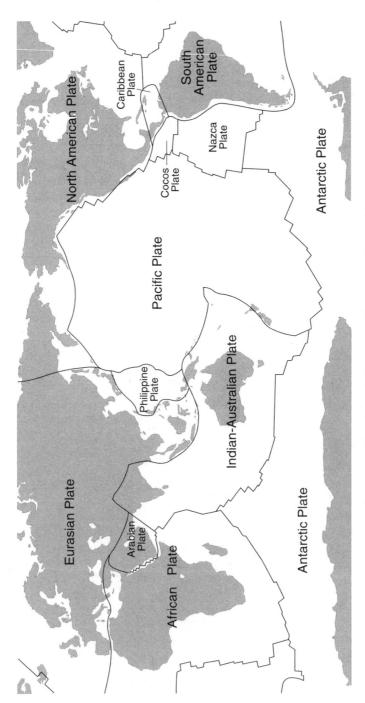

Figure 3-3 Plate tectonic boundaries throughout the world.

The best example of an *accretionary plate boundary* is the ridge that bisects the Atlantic Ocean. New ocean floor is continually being constructed along the crest of this Mid-Atlantic Ridge. Segments on either side of the crest move apart as new molten material is injected into the gap, where it solidifies and then is pushed symmetrically away from the crest. As the material cools beneath the Curie point (the temperature at or below which it can retain magnetization), it assumes the current direction of the earth's magnetic field. The earth's magnetic field has randomly reversed direction in the geological past. This pattern of reversals can be mapped with magnetic observations, and the spreading rate (typically on the order of centimeters/year) away from the axis of the ridge can be deduced. Shallow-focus earthquakes are associated with the divergent or differential movement along this type of plate boundary.

A *subduction* boundary is one where a thinner oceanic plate converges with a thicker continental plate. The sequence of events is illustrated in Figure 3-4. Because the continental lithosphere plate is thicker than the oceanic plate, the net result is that the oceanic plate is forced, or subducted, beneath the continental plate forming a **subduction zone**. Where it turns downward it produces an oceanic trench marking the boundary between the two plates.

As subduction continues, intermediate- and deep-focus earthquakes occur along the boundaries or within the interior of the subducting plate. This zone of earth-

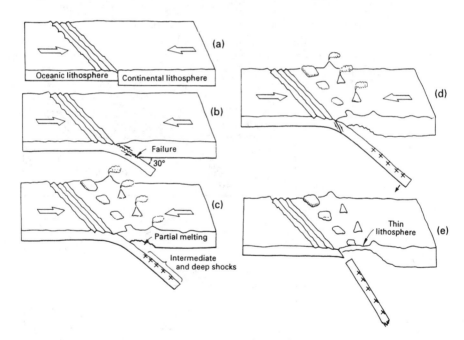

Figure 3-4   A depiction of subduction along a convergent plate boundary. When a thinner oceanic lithospheric plate meets a thicker continental plate, the thinner plate is subducted beneath the thicker plate. The earthquake foci shown on the subducting plate represent the Benioff zone.

quake activity, named the **Benioff zone**, can extend to depths of 700 km or so, with a depth progression that increases beneath the continental side of the subduction zone. As the subducting plate is pushed or pulled downward into a zone of higher temperatures, partial melting can occur and molten material or magma can rise to the surface along zones of weakness. The result is a zone of volcanoes parallel to the oceanic trench, but always on the continental side of the subduction zone.

In the final outcome of subduction the colder oceanic plate is forced farther downward until it breaks off and sinks into the asthenosphere (Figure 3-4e). Most of the world's earthquakes and seismic energy release take place in the subduction zones surrounding the Pacific Ocean.

At a *collision boundary* two continental blocks move together, since the ocean floor between them has been destroyed. As these plates of comparable thickness converge, the one that passes underneath is too light to sink into the asthenosphere, so it slides between the asthenosphere and the other continental plate, raising the latter to great heights. The result is a mountain range (Figure 3-5). The best example of a continental collision boundary is the convergence of the Indo-Australian plate with the continental Eurasian plate that produced the Himalayas, a chain of mountains that first curve away from one another, and then close in to form a knot of massed peaks of extraordinary elevations known as the "Roof of the World." This collision boundary also created a wide zone of deformation and seismic activity where many intermediate-focus earthquakes occur, such as the great 1897 earthquake in Assam, India.

In the final type of major plate boundary, the *transform boundary* , plates slide past each other, neither gaining nor losing material in the process. A classic example is the San Andreas fault transform boundary in California (Figure 3-6). Here the

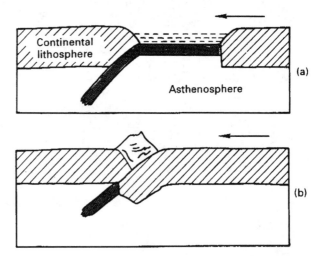

Figure 3-5 A collision boundary between two continental lithospheric plates. The interleaving oceanic lithosphere has been destroyed, resulting in the collision of two plates of comparable thickness. The end result is a mountain range.

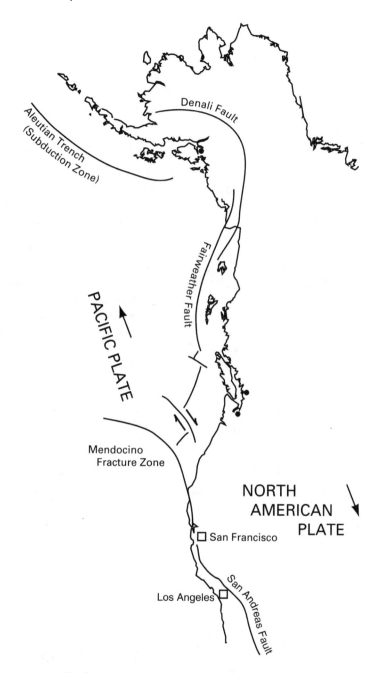

Figure 3-6   The San Andreas fault transform boundary. The Pacific plate is moving northwest along this boundary relative to the American plate.

Pacific plate is moving to the northwest at a rate of 6 cm/yr relative to the North American plate. Los Angeles and San Francisco are now located on opposite sides of the San Andreas transform. At the current rate of movement, Los Angeles will become a suburb of San Francisco 10 million years from now, and 40 million years from now, Los Angeles may well be in the vicinity of the Aleutian Trench.

3-3 EARTHQUAKE EFFECTS

We can classify an earthquake's effects as primary, transient, or secondary, although the rigorous separation of effects is sometimes not possible because they are commonly documented after an earthquake has occurred rather than during it. Nevertheless, *primary effects* are those permanent features produced by the earthquake—surface ruptures, scarps, elevation changes and movement, and offsets of natural and human-made objects. Primary effects lead to *transient effects*—mainly shaking but also tsunami and sandblows. Transient effects cause *secondary effects,* such as landslides, slumps, lurches, damage to buildings, chimneys, and windows, and secondary fires. Table 3-1 summarizes the most common earthquake effects.

One spectacular primary earthquake effect is a horizontal and/or vertical rupture or offset on the surface of the earth that extends for some distance. These ruptures, along zones of weakness called *faults* leave an indelible geological record of the sudden movement. When the rupture is primarily horizontal it is a **horizontal fault offset**, and when it is primarily vertical and breaks the earth's surface it is known as a **scarp**. There are well-preserved scarps in the arid regions of the western United States. Particularly beautiful examples can be seen in the Fairview Peak and Dixie Valley areas, the most seismically active region in western Nevada (Figure 3-7). Two earthquakes in the Dixie Valley–Fairview Peak area in 1954 generated the stunning fault scarp shown in Figure 3-8.

Violent shaking of the ground during an earthquake can produce remarkable changes or shifts in water-bearing strata at depth. During the 1906 San Francisco and 1989 Loma Prieta earthquakes, fountains of water shot up into the air, bringing with them sand from depths of 25 m or so and leaving behind on the surface sand craterlets or sand boils (Figures 3-9 and 3-10). H. C. Gordon describes the earthquake fountains he saw during the Bihar-Nepal earthquake of 1934:

> As the rocking ceased, water spouts, hundreds of them throwing up water and sand, were to be observed on the whole face of the country, the sand forming miniature volcanoes, whilst the water spouted out of the craters, some of the spouts were quite 5 feet high. In a few minutes—as far as the eye could see—was vast expanse of sand and water, water and sand. The road spouted water, and wide openings were to be seen across it ahead of me, then under me, and my car sank, while the water and sand bubbled and spat, and sucked, till my axles were covered. "Abandon ship" was quickly obeyed, and my man and I stepped into knee deep hot water and sand and made for shore. It was a surprisingly cold afternoon and to step into such temperature was surprising.

Table 3-1 Common Earthquake Effects

	Primary	Transient	Secondary
Ground terrain	Surface ruptures, scarps, horizontal offsets, fissures, mole tracks, pressure ridges. Elevation or depression of coastline.	Visible ground waves? Shaking.	Landslides, slumps, mudflows, avalanches, lurches. Sand craters.
Water	Damming of rivers, waterfalls, changes of course. Sag ponds. Permanent water-level changes in wells and springs.	Water-level oscillations in wells, earthquake fountains. Sandblows, seiches, tsunami, and seaquakes.	
Construction	Offsets of fences, roads, ditches, buildings. Bent pipelines and twisted railways.	Creaking, groaning, and swaying.	Common damage to buildings, chimneys, and windows.
Objects		Shaking, rattling, rolling, swinging, and twisting.	Displacement, overturning, falling from shelves.
Potpourri		Nausea and panic. Sleepers aroused; animals disturbed. Trees shake, bells ring, automobiles rock. Sounds, flashes of light.	Clocks stop or change rate. Underwater cables break.

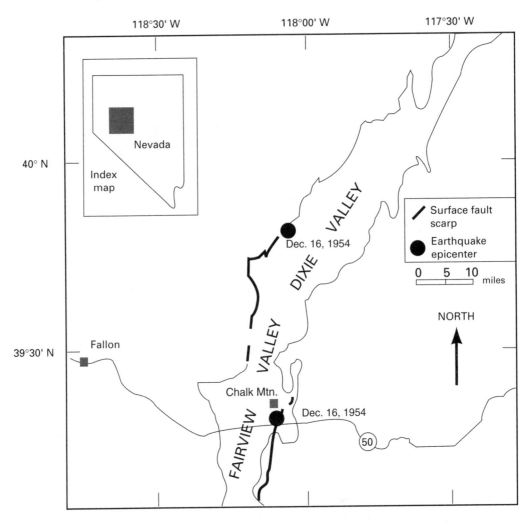

Figure 3-7   Location map of 1954 earthquakes and associated surface ruptures in northwestern Nevada.

One of the most interesting earthquake effects is **brontides**. These earthquake noises should not be confused with miscellaneous noises produced by the agitation of the shocks such as the rattling of windows or shuddering of buildings; rather they are natural sound effects.

Brontides have been frequently documented, particularly in India, where strange sounds are referred to as "Barisal" or "Purnea guns." A. E. English, recorded his experience of earthquake noises during the great Assam, India, earthquake of June 12, 1897:

Figure 3-8 View of the fault scarp produced on the east flank of Fairview Peak by the 1954 Dixie Valley earthquakes. The prominent feature in the right background is Chalk Mountain. (Photograph by Bob Wallace, U.S. Geological Survey.)

Date, from looking at my diary, it must have been the 12th June: time, between 5 and 6 P.M. On return from a stroll after game, at jungle camp on bank of "Theingale," which is some 7 miles south-west of Kyouko village, and about 19 miles of Tagaung, I noticed the water in the tank, which was an old river course containing about 300 yards of shallow water, lapping up against the bank below my tent. My hunters said it must be elephants bathing, but on looking there were none at the other end. Some one then pointed to the trees shaking, and we knew it must be an earthquake. To the best of my memory, about 1/2 hour or less after this I heard one or two distinct booms like cannon shots, and came out of my tent and joined my hunters and followers who had also come out and were listening and counting the booms. We counted about 25 distinct booms at intervals of about 3 or 4 seconds. The booms were about as loud and distinct as a shot from a 7 lb gun turned away from one at, say, four miles or more, if the country was very flat. The sounds were not at

Figure 3-9 Sand craterlets near Watsonville, California, produced by the 1906 San Francisco earthquake.   (Photograph courtesy Department of Special Collections, Stanford University Libraries.)

all like the rumbling of earthquakes I have heard before, and quite deceived me and my hunters, who were all especially intelligent in jungle matters and noises.

Observers have generally characterized brontides as low rumbles that gradually increase in loudness, but are clearly distinct from most thunderclaps. A scientist who was recording an earthquake swarm in 1979 in the Mojave Desert of California described associated earthquake noises as "similar to a tympani note, sharp onset, lower in frequency content than thunder . . . and of short duration." The phenomenon has not been studied in detail, but there is reason to suspect that shallow earthquakes can produce audible airwaves.

Another striking earthquake effect that is well documented is *earthquake lights*. Many instances of luminous phenomena accompanying earthquakes were reported by Japanese investigators from 1930 to the mid-1960s. They have also been reported in the United States, Turkey, and elsewhere. Often these lights are described as radiating

Figure 3-10 Sand boils at Moss Landing State Beach, Moss Landing, California, as a result of the October 17, 1989, Loma Prieta earthquake. The illustration vividly shows the effects of liquefied sands reaching the surface. (Photograph courtesy Dan Orange, University of California at Santa Cruz.)

from a point on the distant horizon, having a blue or bluish color, and being brighter than moonlight. Why there should be luminescence concurrent with or after an earthquake shock is still unknown, but one likely cause is a severe low-level atmospheric oscillation in regions of high electrical gradient. Another possibility is that changing stresses in crustal rocks due to stress buildup and release produce an electromagnetic (piezoelectric) effect. Luminescence before an earthquake, which has been reported occasionally, is even more intriguing.

Horizontal offset is the most visual memorial of an earthquake. By *horizontal offset* we mean that an observer standing on one side of a fault plane and looking across to the opposite side would see a misalignment of features that straddle the fault zone. The most famous examples of horizontal faulting were produced by the San

Figure 3-11   Offset of 8 ft in a fence crossing the San Andreas fault on the Folger Ranch, near Woodside, California, as a result of the 1906 earthquake. Observe the clear right-lateral offset. (Photograph courtesy Department of Special Collections, Stanford University Libraries.)

Francisco earthquake of April 18, 1906. This earthquake left many right-lateral offset fences. *Right-lateral* offset means that if an observer stands on one side of the trace of the fault, the opposite side is seen to have moved to the right. The concept is unequivocal: If the observer steps across the trace of the fault and looks over from the other side, the opposite side is still seen to have moved to the right. Some offset fences observed after the 1906 earthquake were unambiguous evidence of the direction of the movement since the fence lines straddled the fault trace (Figure 3-11). Measurements of these kinds of postearthquake offsets are extremely useful for inferring the average offset or **slip** that occurred on the fault plane during the earthquake. As we shall see later slip is a key parameter for assessing the "true" size of an earthquake.

When there is horizontal movement through heavily alluviated or cultivated terrain, the resulting offset produces a zone or line of disturbance resembling the marks

Figure 3-12   Mole tracks in Mudurnu Valley, Turkey, from an earthquake that took place on July 22, 1967. (Photograph courtesy Bob Wallace, U.S. Geological Survey.)

left by a farmer's plough or the track of a large mole. Figure 3-12 shows a *mole track* formed by an earthquake in Mudurnu Valley, Turkey, on July 22, 1967.

3-4 SEISMIC INTENSITY

Before instrumental observations were possible, it was realized that the effects of an earthquake could be classified according to the apparent intensity of its tremors. Every earthquake has many intensities, the strongest usually being at the epicenter; intensities decrease the farther a region is from this point. Intensities are quantified

Table 3-2 Modified Mercalli Intensity Scale

I Not felt except by a few people under especially favorable conditions. Best considered as an instrumental shock— that is, one noted by seismic instruments only.

II Felt only by people at rest in places such as upper floors of buildings or by sensitive and nervous persons. Delicately suspended objects may swing.

III Felt by many people in places such as upper floors of buildings, but to such a slight degree that most do not recognize it as an earthquake. Standing automobiles may rock slightly, as if from vibrations caused by a passing truck. Duration may be measured. People realize there was an earthquake if others relate they, too, had felt it.

IV In daytime, felt by many indoors but by only a few outdoors. Dishes, windows, and doors are disturbed and walls creak. Sensation is like that of a heavy truck striking a building. Windows rattle and standing automobiles rock considerably.

V Felt by all; many awakened. Some dishes and window glasses break, some wall plaster may crack. Unstable objects are overturned. Disturbance of telephone poles, trees, and other tall objects sometimes noticed. Pendulum clocks stop and doors swing closed and open.

VI People are frightened and run outdoors. They walk unsteadily and are characteristically alarmed. Heavy furniture may move; some instances of fallen plaster and toppling of chimneys. Church bells ring. Damage is generally slight.

VII Everybody runs outdoors. People find it difficult to stand. Damage is negligible in buildings of good design and construction, slight to moderate in ordinary structures, and considerable in poorly built or badly designed structures. Chimneys break. Tremor is felt in moving automobiles. Concrete irrigation ditches are damaged.

VIII Some damage occurs even in buildings of good design and construction; there is great damage in poorly constructed buildings. Panel walls are thrown out of frame structures and chimneys fall. Sand and mud are ejected in small amounts. Tremor hinders driving of automobiles.

IX Damage to buildings of good design and construction is considerable. Structures are thrown out of alignment with their foundations. Ground cracks conspicuously. Underground pipes are damaged. There is fissuring in alluvial ground. Landslides occur.

X Wooden houses of good design and construction collapse. Most masonry and frame structures are destroyed; foundations slide. Water surface rises.

XI Almost all masonry structures collapse. Bridges are destroyed. Fissures appear over entire surface of ground. Underground pipelines go out of service. Earth slumps and slips in soft ground. Rails are prominently bent.

XII Total damage. Waves are seen transmitted at ground surface. Topography changes. Objects are thrown into air.

using the Modified Mercalli Intensity Scale, which has numerical values ranging from I (virtually indiscernible without instruments) to XII (causing total destruction of most buildings) (Table 3-2). After the numbers are assessed at various localities, a contour map, known as an **isoseismal map**, is prepared depicting regions of equal seismic intensity.

Destructive earthquakes are mentioned in the Bible, in the Talmudic writings, and in old Arabic catalogs for the eastern Mediterranean region. Earthquakes in the Holy Land have occurred intermittently since at least 1250 B.C., and possibly as far back as the destruction of Sodom and Gommorrah (2150 B.C.?). There is a damaging earthquake in this region about once every 100 years. On January 1, 1837, a devastating earthquake struck the town of Safed, Israel. A certain Dr. Thomson, who was then in Beirut, Lebanon, described his experiences:

> It was just before sunset . . . when the house began to shake fearfully. . . . *Hezzy! Hezzy!* (Earthquake! Earthquake!). . . . On the morning of the 18th we reached Safed . . . and I then understood for the first time what desolation . . . when he ariseth to shake terribly the earth.

It is estimated that 5000 people in Safed were killed in this disaster, out of a total population of 9000.

Figure 3-13 shows isoseismal maps for two destructive earthquakes in Israel—the earthquake of 1837 centered near Safed and the 1927 one near Jericho—superimposed on a current seismic zoning map for Jordan produced by the Royal Scientific Society. The isoseismal contours for these two events describe their seismic intensity distribution. Note that these contours are elongated ellipses. If there were complete symmetry between the underground structure around the earthquake's focus and the surface structure around the epicenter that lies directly above, the shape of the isoseismal contours would be circular rather than elliptical and the epicenters would be in the very middle of the circles.

There are several problems in assessing earthquake intensity. Intensity maps for earthquakes can be influenced by local geologic effects (such as liquefaction) and bias toward inhabited regions. Most problematical, perhaps, is the reliability of subjective descriptions. In some regions of the United States, for instance, people are so proud of living through an earthquake that they are prone to exaggerate its effects. On the other hand, there are regions of the world where people are embarrassed by their experience or stoical and therefore inclined to play down an earthquake's effects.

There are several misuses of seismic intensity measures. The most common is to correlate them with earthquake size and amplitude of ground motion. Their usefulness for engineering design purposes has also been questioned. For all their shortcomings, however, seismic intensity scales and isoseismal maps can give us a general description of the pattern of earthquake activity and help us detect localized areas of high damage potential, such as landfills, unstable hillsides, and sites capable of liquefaction. (The biblical comparison between the wise man who founded his house upon rock and the foolish man who built his on sand—and great was the fall thereof—still holds!) Seismic intensity studies are also useful for detecting zones of weakness, such as faults. They are certainly valuable for cataloging past earthquakes in areas with a long historical record, such as China and the Middle East.

3-5 EARTHQUAKE MAGNITUDE, ENERGY, AND RECURRENCE INTERVAL

Is there any way to measure an earthquake's size irrespective of its intensity or effects? We have seen that intensity involves a subjective human assessment of shaking produced by an earthquake at a particular point, that is usually highest close to the epicen-

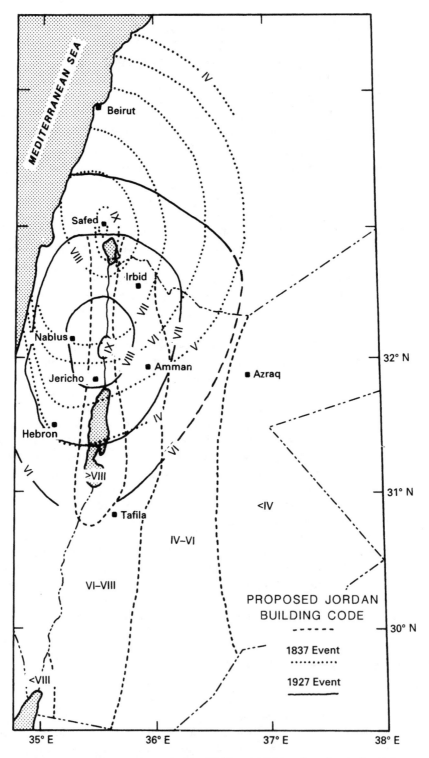

Figure 3-13 Isoseismal maps of the 1837 and 1927 earthquakes in the Holy Land superimposed on a zoning map for a current building code. (From Kovach, 1988.)

ter, and that it depends strongly on site conditions. Suppose that two earthquakes occur tomorrow, one in a densely populated metropolitan area and the other in the middle of a desolate desert. For the first, the amount of damage to structures would be easy to assess and there would be many reports on the degree of shaking felt. The second would probably go unnoticed except by seismologists (*seismology* is the scientific study of earthquakes). How, then, do we describe the size of both of these earthquakes?

Magnitude is a measure that is applicable to all earthquakes, wherever they occur and whatever their depth. It rates an earthquake as a whole, independent of the effects at any particular point. The magnitude scale is based on the simple premise that if two earthquakes occurred at the same place and were recorded on a seismograph located at the same station, the larger earthquake would produce bigger amplitude seismograms. The **seismograph**, an instrument designed to record earthquake disturbances, registers the maximum deflection of seismic waves from their average value—that is, their **amplitude**. The magnitude of our two hypothetical earthquakes, then, would be determined by the maximum amplitude of their seismic waves as seen on the seismogram. The inventor of the magnitude scale, Charles Richter, proceeded in the following manner. He measured the maximum amplitudes observed on seismograms from various earthquakes at different distances out to 600 km. To accommodate the large range of values in observed amplitudes, Richter introduced a *logarithmic scale* to the base 10—meaning each whole-number step in the scale represents a tenfold increase in measured amplitude. He noted that when logarithms of the observed maximum trace amplitudes on the seismograms were plotted against distance, there was a general decrease in amplitude with distance. The relation

$$M = \log A - \log A_0$$

defines the Richter local magnitude (M). A is the recorded trace amplitude for a given earthquake at a given distance, as observed with a particular instrument. A_0 is the amplitude of the mildest tremor that could be detected instrumentally, which Richter took to be 0.001 mm at a distance of 100 km. To this standard he assigned a magnitude of zero. The purpose of $\log A_0$ was to make it possible to assign magnitudes of positive values to the smallest shocks that might be studied; it can be considered a distance correction factor for comparing earthquakes recorded at different distances. As noted, every whole number in the Richter Scale represents a tenfold increase in amplitudes on the seismograms recorded at the same station. Thus a magnitude 8 earthquake has an amplitude 10 times greater than a magnitude 7 earthquake, 100 times greater than a magnitude 6 earthquake, 1000 times greater than a magnitude 5 earthquake, and so on. There is no mathematical bottom or ceiling to the Richter Scale. Apparently, however, earthquakes having a Richter magnitude greater than about 9 do not occur.

Here are the recorded magnitudes for some notable earthquakes of the twentieth century:

Magnitude	Site	Date
5.3	San Francisco	March 22, 1957
5.8	Bakersfield	August 22, 1952
6.3	Long Beach	March 10, 1933
6.6	San Fernando	February 9, 1971
7.1	Nevada	December 16, 1954
7.5	Southern California	June 28, 1992
7.7	Arvin-Tehachapi	July 21, 1952
8.3	San Francisco	April 18, 1906
8.6	Tibet	August 15, 1950

We classify earthquakes with a magnitude greater than 7.5 as *great* and those with magnitudes ranging from 6.5 to 7.5 as *major*. *Large* earthquakes lie in the magnitude range from 5.5 to 6.5. A magnitude of 5.5 is considered to be the threshold of damage for most communities.

It needs to be emphasized that the Richter Scale is based on arbitrary choices, including the type of reference seismograph used, logarithms to the base 10, and the definition of the "standard" earthquake. Nevertheless, it is used worldwide and a number from this magnitude scale is what a seismologist would tell you if you asked how large a particular earthquake was. However, *intensity* is the parameter that interests engineers responsible for designing earthquake resistant structures, for that is what counts in terms of earthquake hazards. A good analogy for relating the two measures is that between the power of a radio station (magnitude) and the strength of the received signal at some distant point (intensity).

In some areas where sufficient instrumentation has been in place long enough to accumulate a substantial record of earthquake effects, attempts have been made to establish crude correlations between earthquake intensities and magnitudes. The following table shows such a correlation:

Magnitude	2	3	4	5	6	7	8
Intensity range	I – II	III	V	VI – VII	VII – VIII	IX–X	XI – II
Radius of perceptibility (km)	0	15	80	150	220	440	800

Earthquake magnitude is useful not only for establishing the size of an earthquake but also for gauging its energy, or capacity to do work. (An earthquake's *work* is related to the displacement of crustal material.) To calibrate an energy scale, we use a known energy source that generates seismic waves: nuclear explosions. When we plot magnitude against energy release, we find that the range of energies involved is so large that, again, we must use a logarithmic scale. Figure 3-14 is a plot of the energy released from nuclear explosions and that of the magnitude 8.3 San Francisco shock of April 18, 1906. Note that the relation between energy and magnitude is empirically given by the relation

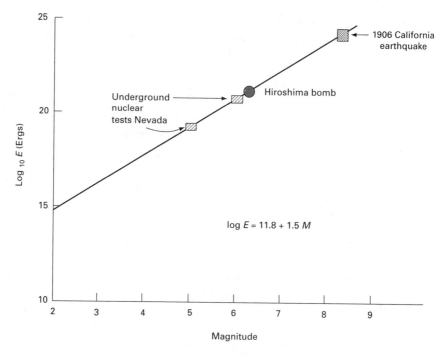

Figure 3-14   Plot of the logarithm of energy release versus magnitude of earthquake of explosion (10^7 ergs = 1 joule).

$$\log E = 11.8 + 1.5M$$

What this relation tells us is that there is roughly a sixfold increase in energy released for each half-step increase in magnitude.

We can now answer a frequently posed question: How many little earthquakes would it take to equal one large shock? It turns out that 31.6 magnitude 7 earthquakes or 1 million magnitude 4 earthquakes would be needed to equal the energy release of a magnitude 8 shock! So you can see how misconceived is the popular idea that many little earthquakes act as a safety valve preventing the occurrence of a large earthquake.

The largest magnitude thus far recorded for an earthquake is about 8.6. Should we except larger-magnitude earthquakes? After all, the time span of instrumental coverage is only about a century—an infinitesimal point on the geological time scale—and larger earthquakes may have occurred in the past. Two lines of reasoning suggest that a Richter magnitude of 9 is about the limit of possibility:

1. An earthquake of about magnitude 10 would correspond to a zone of failure that completely circles the earth. We see no evidence for continuous zones of failure of this length.
2. By studying the statistics of smaller events, we can determine how many years should elapse between magnitude 9.9 earthquakes. As we will see the recur-

rence time is such that we should have historical, if not instrumental records of such an event. There is no such record, and it is highly unlikely that such a catastrophe would go undescribed.

Before we determine the recurrence interval of a magnitude 9.9 earthquake, we need to digress a bit into earthquake statistics. We are concerned with how many earthquakes of specified magnitude M or greater occur annually. The following table gives the average number of earthquakes throughout the world per year in the different magnitude ranges:

Magnitude	5	6	7	7.9	8.6	
Number	2000	180	17	~2	0.25	Shallow focus
			30	6	0.2	Intermediate focus
			10	2	0.1	Deep focus

Notice that for every magnitude there are many more shallow-focus earthquakes than intermediate- or deep-focus shocks. This is quite pronounced for very large magnitude earthquakes. As we look across rows we also see that small earthquakes are much more numerous than large ones. We can use this information to infer the number of earthquakes expected per year of any specified magnitude. In other words, the expected recurrence interval or return time for a particular magnitude earthquake can be calculated. (This will be very important when we pursue the question of earthquake risk in Appendix D.)

Plotting the logarithm of the cumulative annual number of earthquakes of specified magnitude M or greater against magnitude gives a relation of the form

$$\log N = a - bM$$

or

$$N = 10^a 10^{-bM}$$

In our case, for shallow focus earthquakes, we can determine that $a = 8.7$ and $b = 1.1$ satisfy our data. This relation allows an extrapolation to infer that a magnitude 9.9 earthquake should occur every 100 years or so—well within the reliable historical record. Since no historical event for which we might reasonably assign such a high magnitude has been documented, we conclude that there has never been a 9.9 magnitude earthquake.

3-6 SEISMIC MOMENT AND SLIP RATES

One of the shortcomings of the earthquake magnitude measure is that the scale effectively saturates for very large earthquakes—that is, very large earthquakes radiate a significant amount of energy at low frequencies that are not easily recorded by conventional seismographs. For this reason seismologists believe that it is more accurate to discuss the size of an earthquake in terms of its seismic moment M_0. **Seismic moment** is equal to the surface area of the fault area being displaced, multiplied by the average displacement distance and the rigidity of the rocks involved.

We can express this as

$$M_0 = \mu DA$$

where M_0 is seismic moment, μ is the rigidity of the rock material, D is the average offset or displacement across the zone of failure, and A is the area of the fault or failure zone. (See Appendix C for a fuller discussion of seismic moment.)

Seismic moments of some important earthquakes are shown below:

	Magnitude	Area (km^2)	Offset (m)	M_0 (dyne-cm)
Truckee (1966)	5.9	100	0.3	8.3×10^{24}
San Fernando (1971)	6.6	280	1.4	1.2×10^{25}
Loma Prieta (1989)	7.1	1,000	1.0	3.0×10^{26}
Alaska (1964)	8.5	150,000	7.0	5.2×10^{29}
Chile (1960)	8.3	160,000	21.0	2.4×10^{30}

NOTE: 1 dyne-cm equals 10^{-7} newton-m.

Examination of this table shows that the 1964 Alaska earthquake had a magnitude of 8.5 compared with a magnitude of 8.3 for the 1960 Chile earthquake. Yet in terms of seismic moment, the Chile event was about five times larger. It is clear why: Seismic moment is governed by the product of fault area and offset, and the Chile earthquake produced a much greater offset or displacement.

The seismic moment of an earthquake is generally extracted from analyses of seismograms. However, when the fault actually breaks the surface, as happened in the 1906 San Francisco earthquake, we can estimate the amount of displacement from the offset of human-made or identifiable geologic features. If the fault does not rupture at the surface, we can estimate the area involved from the spatial distribution of the **aftershocks**—smaller earthquakes that follow the main event.

One important application of seismic moment is that it allows us to determine a seismic **slip rate**—the rate of movement between lithospheric plate boundaries—and compare this to the long-term geological slip rate inferred from plate tectonics. Regions along the plate boundary where the slip rate has not kept pace with the geological slip rate are likely candidates for a future great or major earthquake, depending on the amount of slip deficiency.

Let us look at the Kurile Islands zone, an active sector of the circum-Pacific seismic belt. Assuming a length of about 500 km and a zone of seismic activity extending to about a depth of 300 km, this sector has a potential fault rupture area of 150,000 km^2 or 1.50×10^{15} cm^2. An examination of the earthquake data for the time interval 1904 to 1980, a span $T = 76$ years, yields a total seismic moment release of

$$\Sigma M_0 = 2.0 \times 10^{29} \text{ dyne-cm}$$

We know that

$$\Sigma M_0 = \mu DA$$

so the seismic slip rate is

$$\frac{D}{T} = \frac{\Sigma M_0}{\mu AT}$$

Using $\mu = 5 \times 10^{11}$ dynes/cm^2, we obtain a seismic slip rate of

$$3.5 \text{ cm/yr}$$

If we know the long-term geological slip rate is actually 10 cm/yr, we see that, in terms of seismic moment, we have a slip deficit of

$$\Sigma M_0 = 5.7 \times 10^{29} \text{ dyne-cm}$$

An earthquake comparable in size to the 1960 Chile or 1964 Alaska earthquake is required for the seismic slip rate in the Kurile Islands zone to catch up with the long-term rate. This is clear demonstration of why major earthquakes tend to occur in those areas of plate boundaries, called **seismic gaps**, that have not been subjected to rupture in the recent historical record. A seismic gap does not mean there has been a cessation of seismic activity in this area, only that the crustal rocks are resisting for a longer period than usual the stresses being placed on them. The longer these stresses accumulate, the more violent the eventual earthquake will be.

3-7 EARTHQUAKE PREDICTION, CONTROL, AND MODIFICATION

Predicting an earthquake means specifying the time, place, magnitude, and probability of its occurrence. Note that a **prediction** is far more specific than a **forecast**, which merely gives the probability that an event—in this case, an earthquake—of a specified size will occur within a considerable time span. Predictions are obviously more valuable than forecasts for safeguarding populations in seismic zones.

Earthquakes cannot yet be predicted with any certainty, though, because we do not fully comprehend the mechanism that produces them. Our knowledge of plate tectonics and sea-floor spreading has grown enormously in recent years, however, and one mega-theory of earthquake prediction has taken advantage of these advances. This theory uses *seismic gaps* to pinpoint likely sites of future major earthquakes. Its premise is that since these gaps have not experienced large shocks for considerable periods, they are logical candidates for future earthquakes. The trouble with this theory is that it is more a forecast than a prediction: It tells us the places where there is a strong probability of earthquakes, but not the time when they will occur. Still, it has encouraging possibilities.

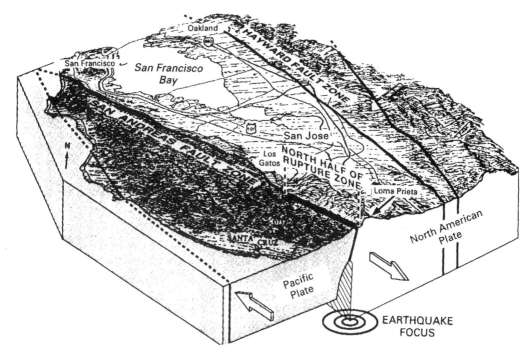

Figure 3-15 Schematic diagram showing location of Loma Prieta earthquake of October 17, 1989, and its relationship to the boundary between the North American plate and the Pacific plate. (Diagram courtesy U.S. Geological Survey.)

On October 17, 1989, a magnitude 7.1 shock that came to be known as the Loma Prieta earthquake shattered the San Andreas fault zone in central California (Figure 3-15). There had been many signs that a large earthquake was brewing in this region. The San Andreas fault forms the main boundary between the Pacific plate and the North American plate, which is moving to the southeast relative to the Pacific plate at an average rate of 5 to 6 cm/yr. Even when slip rate is constant over millions of years, there are times when movement between plate boundaries is sudden and violent, and that is when large earthquakes occur along those regions of the boundaries we call "seismic gaps." The Loma Prieta earthquake took place in one such gap, as a portion of the plate boundary that had been temporarily "locked" suddenly snapped under the continual driving motions of the lithospheric plates.

Figure 3-16 is a vertical cross section of seismic activity from north of San Francisco to south of Parkfield, California, showing the location of earthquakes from January 1969 through July 1989. Notice that there are three zones of sparse seismic activity: the Parkfield gap, the southern Santa Cruz Mountains gap, and the San Francisco Peninsula gap. The lower vertical section of the figure shows the Loma Prieta earthquake and its aftershocks filling the Santa Cruz gap. Although this earth-

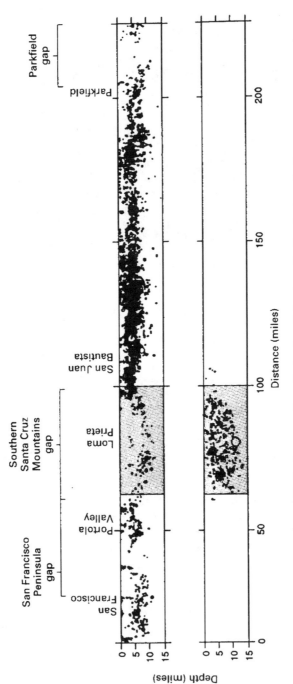

Figure 3-16 Vertical cross section of seismic activity along the San Andreas fault zone from north of San Francisco, southeastward to Parkfield for the period from January 1969 through July 1989. Observe that the subsequent Loma Prieta earthquake and its associated aftershocks filled a gap in earlier seismic activity. (Diagram courtesy U.S. Geological Survey.)

quake had not been "predicted" in the full sense (a definite time had not been specified), its occurrence did lend credence to the seismic gap concept of earthquake prediction. It is now believed that we can assign probabilities to future major earthquakes along the San Andreas and related faults within a 30-year time frame.

Another prediction scheme devised in recent years relies on *premonitory phenomena*. It has been observed that there are certain physical precursors to large earthquakes. Among these signs are tilting of the ground, local deformations of the earth's
crust, variations in the electrical resistance of rocks, changes in the water levels of
wells, and increases in the number of smaller seismic events. Tilting of the ground,
especially, has been found to correlate well with a general increase in seismic activity.
For instance, anomalous tilt was observed in 1966 just prior to some earthquakes in
Matsushiro, Japan (Figure 3-17). Tilting is monitored using a long-baseline horizontal tube containing water: When the ground tilts, the positions of the tube will change,
but the fluid will seek the level, or reference, position.

The premonitory method of prediction has proved reasonably successful in
Japan and China, where the most ambitious observational studies have been under-
taken. In some areas the authorities have been able to estimate the location and magnitude of severe earthquakes several months beforehand and issue timely warnings.
Thus the magnitude 7.3 Haicheng, China, earthquake of February 1975 was predicted,
and a million people were evacuated before it struck. This was the first time a major
earthquake had been predicted anywhere, and seismologists all over the world were
encouraged. But the magnitude 7.8 Tangshan, China, earthquake of 1976 was *not*
foreseen, and there were at least 250,000 fatalities.

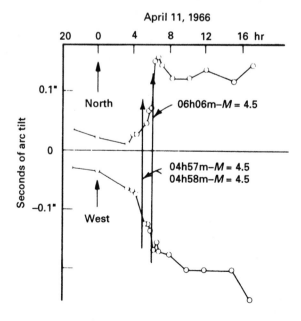

Figure 3-17   Ground tilting observed
prior to some earthquakes that
occurred in the vicinity of Matsushiro,
Japan.

A third approach to earthquake prediction is based on the hypothesis that a great earthquake—that is, one of magnitude 7.5 or higher—will be preceded by an increase in small and medium-sized shocks in a broad region (a seismic gap?) that includes the site of the anticipated great shock. A fairly sophisticated computer algorithm is being tested to identify *times of increased probability* (TIPs) of earthquake occurrence in various parts of the circum-Pacific belt. It could be argued that the TIPs program is more in the nature of forecast than prediction, yet it is based on a thesis generally believed to be valid and is still in the early stages of testing. If TIPs proves successful and capable of refinement, it may one day reach the confidence level of prediction.

We now turn to the subject of earthquake modification and control. After the disastrous San Francisco earthquake of 1906, David Starr Jordan, then president of Stanford University, wrote: "No one can check an earthquake or modify its action. Fortunately, also no one can set it off or stimulate it to any greater violence than nature has intended." This is no longer true. Today we know that human beings *can* trigger earthquakes, and our experience in doing so inadvertently has yielded some intriguing possibilities for more deliberate action.

The filling of Lake Mead behind the Hoover Dam on the Colorado River in 1935 was followed by more than 600 local tremors in the next decade. This marks the first time that human action set off earthquakes—at least that we know of. Since then, there have been other notable examples of human-triggered earthquakes, some with deadly results.

In Kremasta, Greece, the filling of a large artificial lake resulted in a magnitude 6.3 earthquake on February 5, 1966. Figure 3-18 gives a clear picture of the association. A far more devastating shock was precipitated the next year in India by the building of a reservoir. What is particularly interesting about this example is that there was no known seismic activity in the area before the impounding of the reservoir

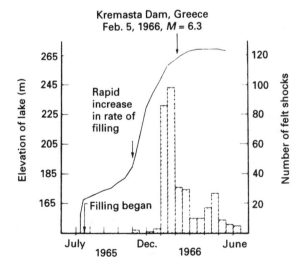

Figure 3-18   The filling of a large lake behind a dam in Kremasta, Greece, precipitated a major earthquake on February 5, 1966.

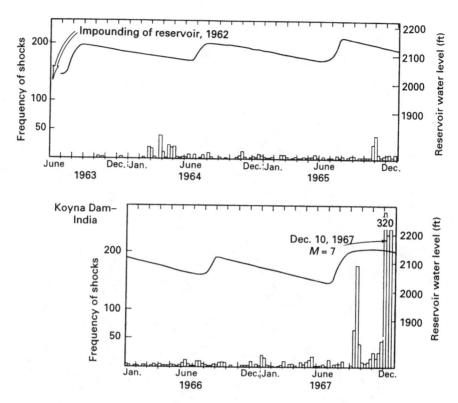

Figure 3-19   The Koyna, India, earthquake of December 10, 1967, occurred as a result of the impounding of a reservoir. The cyclic loading of water and its retention at high levels from August 1967 to December 1967 was responsible for a disastrous earthquake.

in 1962. Indeed, this whole area—the Deccan shield—was believed to be very non-seismic. However, after water was retained in the reservoir above the 2140-ft level from August to October 1965, high seismic activity was noted in November of that year (Figure 3-19). An even higher level of water was retained in the reservoir from August to December 1967, and on December 10 the Koyna earthquake wreaked widespread destruction and cost 200 people their lives.

One last example of human-made earthquakes: The Rocky Mountain Arsenal Disposal Well near Denver, Colorado, was drilled to a depth of 2671 m into crystalline rocks for purposes of disposing radioactive waste fluids. After injection of the fluids was begun on March 8, 1962, a very striking correlation was noted between the amount of fluid injected and the number of earthquakes in the vicinity (Figure 3-20). Termination of the injections in February 1966 did not end the earthquakes: A magnitude 5.0 shock occurred on April 10, 1967, a 5.5 shock on August 9, and a 5.1 shock on November 26 (Figure 3-21).

Intensive study by a U.S. Geological Survey group revealed a strong correlation

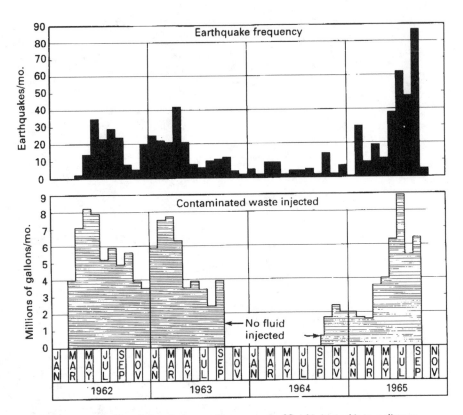

Figure 3-20 Correlation between the amount of fluid injected into a disposal well near Denver, Colorado, and earthquake activity. (Diagram courtesy J. H. Healy, U.S. Geological Survey.)

between fluid pressure at the bottom of the well and the level of seismic activity between 1962 and 1966, but the increase in seismic activity *after* injection was ended ruled out any simple, direct relationship (Figure 3-22). There is, however, quite a satisfactory explanation for the post-1966 earthquakes. Fractures already existed in the rocks when fluid was injected into the well. The injection rapidly increased pressure near the well, reducing the rocks' resistance to fracturing and causing cracks to propagate until they encountered a stronger "lock point"— that is, a place where normal stress across incipient fracture planes was high enough to halt the fracturing. After the injections were stopped, there was a rapid decrease in pressure near the well itself, but the pressure front continued to advance away from the well. The larger cracks that propagated out beyond the pressure front during injection were reactivated, whereas the shorter cracks near the well in which fluid pressure was decreasing became inactive.

All of these experiences with earthquakes triggered by fluids pumped between unstable rock layers raise some provocative questions: What role does the natural infiltration of water into the lithosphere play in producing earthquakes? Can we avert

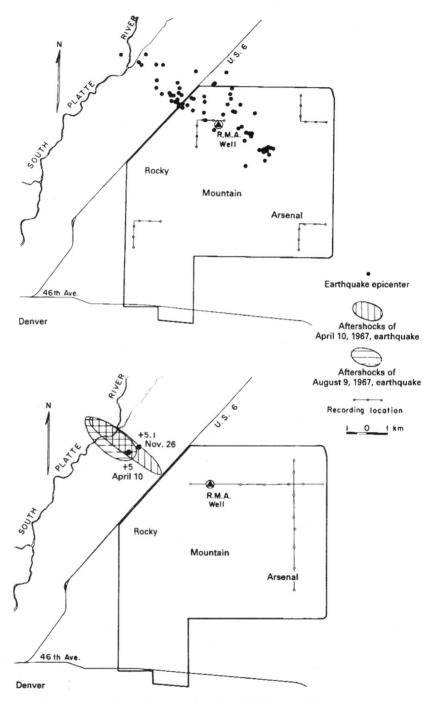

Figure 3-21 Location of the Rocky Mountain Arsenal disposal well relative to Denver, Colorado, and the locations of triggered earthquakes. Note the April 1967 and August 1967 earthquakes occurred *after* the injection of radioactive waste fluids was ended in February 1966. (Map courtesy J. H. Healy, U.S. Geological Survey.)

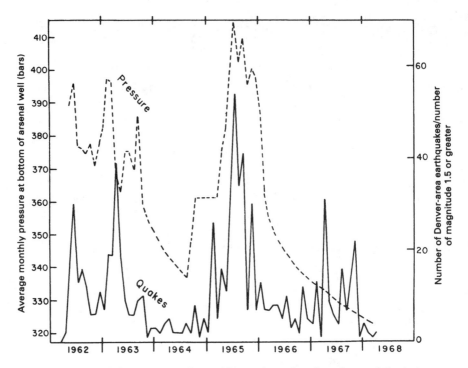

Figure 3-22 Comparison of monthly number of earthquakes and the bottom-hole pressure at the Rocky Mountain Arsenal Disposal Well. (Diagram courtesy of J. H. Healy, U.S. Geological Survey.)

a large earthquake by injecting water into the crust in highly seismic areas to release pressure in smaller earthquakes? Can we use widespread well pumping to choose the time an earthquake judged to be inevitable will occur? *Should* we do any of these things?

Right now our best defenses against earthquakes are rapid warning systems and earthquake-resistant buildings. The second defense is the subject of our next chapter.

REVIEW

1. Where do 80% of the world's earthquakes occur? What smaller region of the world contains 15% of earthquake activity in terms of energy release?
2. Distinguish between an earthquake's epicenter and its hypocenter (focus).
3. Distinguish between lithosphere and the asthenosphere. In which of these layers do earthquakes occur? Seismic creep?

4. Why do shallow focus earthquakes occur along oceanic ridges (accretionary plate boundaries)?

5. Where are intermediate- and deep-focus earthquakes commonly located?

6. What is the difference between a scarp and a horizontal fault offset?

7. What is a brontide?

8. What is an isoseismal map?

9. What are the three major problems in assessing earthquake intensity?

10. Your friend tells you, "That earthquake got me out of a deep sleep. Some plaster fell on my head and my dresser moved across the room." From his statement you know that the *Modified Mercalli Intensity* was at least what value?

11. We want to estimate the magnitude of an earthquake required to completely circle the globe. We make some simplifying assumptions concerning storage of the energy released in an earthquake. Assume that the energy is stored in a box surrounding the fault.

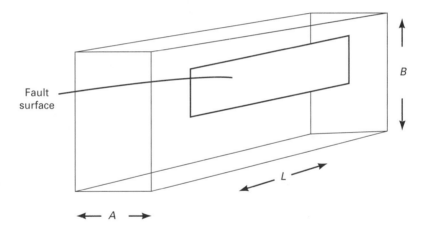

The total energy released is approximately $E = KABL$ where K represents the energy stored per unit volume. Assume that large earthquakes differ in energy release only through the fault length L (i.e., assume K, A, and B are constant for large earthquakes). Knowing that (1) a magnitude 8.5 earthquake corresponds to $L = 400$ km (log 400 = 2.60), (2) log $E = 11.8 + 1.5M$, (3) the circumference of the earth is 40,000 km (log 40,000 = 4.60), and (4) log $E = \log L + \log (KAB)$, what magnitude earthquake would be required to produce a fault with L equal to the circumference of the earth?

12. The magnitude of the 1971 San Fernando earthquake was 6.4, and the 1906 San Francisco earthquake reached an estimated magnitude of 8.3. By approximately what factor was the ground motion at San Francisco greater than that at San Fernando?

13. Describe three theories of earthquake prediction.

4

Earthquake Engineering

According to the design blueprints for our building, there was no cause to fear an earthquake. Our complacency was quickly shattered that miserable rainy day when the earthquake struck. We could hardly believe how violently our massive concrete structure was being shaken. Bookcases were ripped from the walls and volumes of papers almost buried my colleagues and me as we hid beneath our desks. There was no mistaking the immense shaking and force for anything but an earthquake—would the building fall down? After 15 seconds or so, we crawled out from beneath our desks and somewhat gingerly groped our way out of the building, hoping that further tremors would not affect our escape route.

4-1 FUNDAMENTAL PRINCIPLES OF EARTHQUAKE ENGINEERING

Good earthquake engineering produces buildings that are either solid and stiff enough to resist disintegration or elastic enough to "give" rather than collapse during the shaking of an earthquake. They follow a simple, regular square or rectangular ground plan. Load-bearing walls are solidly attached to a substantial foundation, and each floor is solidly affixed to the load-bearing walls. They are erected on suitable ground, preferably hard bedrock and not soft-soil deposits, such as artificial landfill. (Soft-soil deposits are responsible for the majority of earthquake damage, and 80% of the loss of life has occurred in these areas.)

The dynamic response of a building subjected to earthquake ground motion is determined by its *natural frequency* or *period of oscillation.* The taller the building, the greater this will be; a rough estimate of the natural period of oscillation for a building can be obtained by counting the number of stories and multiplying by 0.1. The effects of earthquake shaking are studied by applying lateral forces that simulate the inertial loads a building is subjected to when it is oscillated, or accelerated from side to side, during an earthquake. The applied design force can be visualized as a push by a giant hand (Figure 4-1).

As a building is violently oscillated, it may undergo drift and internal structural failure. Figure 4-1 shows the phenomenon of **drift**, which is the maximum deflection from the vertical of the top of a building during shaking. It produces distortion of windows, interior walls, elevator shafts, and service ducts. It may also cause beams to buckle and fail, columns to compress and collapse, and walls to shear (slide relative to each other in a direction parallel to their points of contact.)

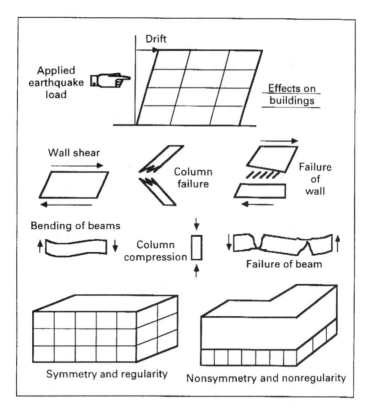

Figure 4-1   Effects on buildings of an applied earthquake load. Drift is defined as the maximum deflection from the vertical of the top of a building. The bottom part of the diagram illustrates the key to good earthquake-resistant design: symmetry and regularity.

The key to earthquake-resistant construction is well-braced walls that are *regular* and *symmetrical* with height (Figure 4-1). Buildings with drastic irregularities in floor plan from story to story will experience greater damage from tremors than those that are regular and symmetrical and made of high-quality earthquake-resistant materials. As might be expected, a building designed to withstand large earthquake loads is significantly more expensive to construct than one designed for smaller loads. There is a definite trade-off here.

Modern architecture from 1930 to 1950 was characterized by either strong steel framing or well-constructed brick masonry. One feature of buildings erected during this era stands out: They had symmetrical and regular profiles. The prevailing architectural philosophy stressed good-quality materials, regularity, and the ability to withstand relatively low inertial loads. In recent decades architects have favored more open and irregular structures that often use prefabricated flexible elements. The net result of these architectural amenities has been a reduction in regularity and stiffness—and many notable building failures during large earthquakes such as the ones in Caracas, Venezuela, in 1967, San Fernando, California, in 1971, and Managua, Nicaragua, in 1972.

4-2 BUILDING CODES

Governments' reaction to all these building collapses was to strengthen building codes in seismic regions. The main emphasis was on implementing proper bracing techniques, setting limits on building drift, and increasing (even doubling or tripling) design code load levels, that is, the level of ground shaking that a building is expected to withstand. The new requirements led to higher costs and delayed new construction and reconstruction, but they did nothing to eliminate or correct bad building configurations or to discourage nonflexible (brittle) construction elements.

What are the essential considerations in drawing up a rational building code? Foremost is seismic risk zoning. Second are importance factors such as what use the building will be put to (office, hospital, school, etc.). A third consideration is setting limits on drift. A rational building code cannot be a panacea but allows for reasonable earthquake-induced ground-shaking levels and recognizes that earthquakes (in the magnitude range of 6 to 7.5) should not produce damage and buildings should not collapse even in catastrophic earthquakes (those with magnitudes greater than 7.5). It needs to be emphasized that codes that are applicable to the amount and type of seismic activity to be expected in a particular region must be adopted.

Seismic risk zoning or *regionalization* data encompass the probability of an earthquake's occurrence, expected intensity and frequency of tremors, the character of the ground subjected to shaking, and the distance to the causative fault. These data have three main audiences: geologists, insurance companies, and engineers. Geologists use the data to augment information they have obtained from epicenter maps, which are often biased by the clutter of many small events, the location of seismic recording stations, and an emphasis on populated areas. Insurance companies use the data on expected intensity and frequency of tremors to estimate risk when setting insurance rates. Engineers focus on expected maximum intensity and/or ground acceleration.

Certain inaccurate assumptions are commonly employed in assessing or constructing seismic regionalization maps based on these data. One is that in an area where small earthquakes are frequent a great earthquake should be anticipated. Often, however, this is not true. Another false assumption is that the strongest level of shaking that has occurred in the past at a given locality is not likely to be exceeded in the future. This supposition ignores the importance of the *nature* of the ground in question in predicting the size of future shaking. For the worst ground, usually found on ocean shores or next to bodies of water, what happened yesterday is a very uncertain norm for what may happen tomorrow.

Microregionalization or microzoning is imperative for land-use planning in seismic regions since it is the only kind of assessment that satisfies engineering requirements in these areas. It is growing practice in the United States, Italy, Japan, Mexico, and several other countries that have suffered damaging earthquakes. What do we mean by *microregionalization*? It is the identification in terms of earthquake hazards of various ground conditions in a city or municipality. The most important factors in making this identification are

- Soil conditions
- Susceptibility to landsliding and land instability
- Topographic variations that enhance ground-shaking events
- Proximity to mapped geological faults

Various types of soils, such as loose, unconsolidated materials and water-saturated earth, can amplify ground motion. Certain terrains are susceptible to ground settlement, landsliding, and **liquefaction**, which is the process whereby a mixture of soil and sand behave like a fluid rather than a wet solid mass. To properly assess the seismic intensity and ground acceleration of future earthquakes, then, the first requirement is detailed geological mapping. Next we need to know the locations of **active faults**—those in which movement has occurred in historical time or that now give evidence of earthquake activity—and the documented pattern of seismicity for the area. With this knowledge we can map, on a detailed local scale, the vulnerability of areas susceptible to anomalous ground-shaking effects.

Once we construct a seismic regionalization map that lets us evaluate local site conditions and quantify the expected level of ground shaking, we need to translate this information into a building code that specifies the **shearing resistance**—the horizontal frictional resistance at the base of a structure opposing the applied horizontal acceleration of a tremor—of all structures to be built on the site. To meet these specifications engineers use the concept of equivalent lateral force or *base shear coefficient* that a building must be designed to withstand. The **base shear coefficient** is related to the percentage of the earth's gravitational **acceleration** that the building will receive during intense ground shaking. (The earth's gravitational acceleration, g, equals 980 cm/sec^2. We call this acceleration 1.0 g.)

The stresses an earthquake puts on a building depend on peak ground acceleration and the duration and frequency of the ground movements. Large-magnitude earthquakes can produce severe horizontal and vertical ground shaking and distortion, along with twisting and rocking. Thus the structural elements of a building are subjected to a complex pattern of forces. The common measure of all these violent motions is *peak ground acceleration*, or the maximum value of acceleration the ground motion achieves. Sometimes this is hard to ascertain because of the complicated directionality of the horizontal motion.

The next critical aspect of ground shaking is its *duration*, or the length of time a given level of acceleration exceeds a certain base level. The higher the level of acceleration and the longer it continues, the more destructive the earthquake will be.

The final critical aspect of strong ground motion is its *frequency*, or the number of times the ground moves forward and backward per second. Earthquake ground motion is rarely sinusoidal; rather, its waveforms contain different frequencies and amplitudes. Once the predominant or driving frequency of strong earth tremors in the area is identified, building design can be adapted to slow the natural frequency of oscillation in structures so that *resonance*—identicality between the natural period of oscillation and the predominant frequency of ground motion—is avoided.

The typical building code specifies that the lateral force coefficient, V_0, for building design must be

$$V_0 = ZIKCSW$$

where

Z = numerical coefficient ≈ 1

I = occupancy factor ≈ 1

S = 1 (depends on period of oscillation of building)

W = dead + live load (the weight)

K = 0.67 to 1.33 (depends on the structural system of the building)

C = $1/15\sqrt{T}$ (does not have to exceed 0.12 where T is the natural period of oscillation of the building in seconds)

It can be demonstrated that $KC = S_A/g$, where S_A represents the maximum ground acceleration as a function of frequency to be accommodated in the building design.

Let us compute the maximum lateral acceleration for a 20-story building. As noted earlier the natural period of oscillation, T, for tall buildings can be estimated from

$$T = 0.1 \times \text{number of stories}$$

Therefore,

$$T = 0.1 \times 20 = 2 \text{ seconds}$$

What does this mean as a measure of the ground acceleration applied to the base of the structure?

$$KC = \frac{S_A}{g}$$

$$K = 0.67 \text{ to } 1.33$$

$$C = 1/15 \qquad = 0.05$$

so that

$$\frac{S_A}{g} = 0.03 \text{ to } 0.07 \quad \text{or} \quad 3\% \text{ to } 7\% \text{ of } g$$

Buildings that are subjected to lateral accelerations greater than these levels from intense ground shaking can deform (sway or bend) beyond their linear elastic capacity and therefore fail or collapse.

Nuclear power plants, for obvious reasons, are subjected to very stringent design levels of ground acceleration in earthquake regions. Building codes specify two design levels of acceleration: one for a *safe-shutdown earthquake*, typically taken to produce ground shaking that is 75% of g, and the other for an *operating base earthquake*, one whose level of ground shaking would not prevent continued operation of the plant without undue risk. A safe-shutdown earthquake is the maximum credible earthquake for the region; it has only a small probability of occurring during the lifetime of the nuclear power plant and has a long recurrence interval. An operating-base earthquake has a much higher probability of occurring and its recurrence interval is about half that of the safe-shutdown earthquake. These specified design levels possess maximum prudence—they go well beyond design standards for other structures.

Many people are opposed to the construction or continuation of nuclear power plants in regions that are at all susceptible to earthquakes. A nuclear power plant had been operating in Humboldt County, California, for several years when a magnitude 5.3 earthquake occurred near the plant. Although there was no structural damage to the plant from this modest-size earthquake, the opposition to nuclear power in the region became so highly charged that the plant was permanently closed down.

4-3 COLLAPSE-PRONE CONSTRUCTION

It is one of the unfortunate facts of life that earthquakes often occur in communities that are least prepared by modern construction standards to withstand them. An engineering axiom has it that earthquakes do not kill, buildings do. About three-quarters of the fatalities attributed to earthquakes are the result of building collapse. Let us look at some examples.

On August 31, 1968, a magnitude 7.3 earthquake in northeastern Iran in the vicinity of Dasht-e Bayāz, some 250 km south of Mashad on the northern border, killed

10,000 people and left another 60,000 homeless. Seismic intensity in the epicentral area was assessed at a Modified Mercalli Intensity of X. The typical building in the area was a one-story structure with a domed roof made of one or two layers of brick covered by a layer of mud. No wood or steel beams were used for reinforcement, and the structures were erected directly on the ground surface with no footings. Figures 4-2 and 4-3 show some of the damage from this earthquake in Kakhk, a village of 8000 people located in the epicentral area. Half of the population perished in the rubble.

Figure 4-2 illustrates a centuries old construction technique that may have evolved in response to repeated earthquakes in the Middle East. Notice that both the building with the spherical-shaped dome and the arches are relatively undamaged. Even though the construction materials used were rather primitive, the architect observed the principles of basic earthquake-resistant design: The structures are symmetrical, regular, and not too large in lateral dimension or elevation. Sinan, a sixteenth-century Armenian architect commissioned by the Ottoman sultan Suleiman the Magnificent, not only used this kind of design in the many mosques that he built but also placed chain reinforcements around the domes and towers.

The concept of a regular building shape and the utilization of a low profile in design has stood the test of time. A powerful earthquake, with a Richter magnitude of

Figure 4-2 Damage in village of Kakhk from the magnitude 7.3 Iranian earthquake of August 31, 1968. Note the survival of the spherical-domed building on the right and the arches in the middle. (Photograph courtesy R. D. Brown, U.S. Geological Survey.)

Figure 4-3   Damage in Kakhk from the Iranian earthquake of August 31, 1968. (Photograph courtesy R. D. Brown, U.S. Geological Survey.)

8.1, struck the island of Guam in the western Pacific Ocean in August 1993. Guam is located about 1600 km east of the Philippines. This earthquake occurred as Guam was being subjected to very heavy rains from Typhoon Steve; thus it could be described as a double sudden-onset natural disaster. Ground shaking lasted for at least 60 seconds and triggered landslides. Surprisingly, there were no reported fatalities. Some damage to buildings was reported, but there was no collapse of any buildings. The lack of building collapse is because most buildings on Guam are constructed with reinforced concrete, are *low in profile,* and are designed to withstand 320 km/hr winds.

Most people associate the year 1906 with the famous San Francisco earthquake of April 18, which had an estimated magnitude of 8.3. In this same year on August 17, a magnitude 8.6 earthquake practically destroyed the city of Valparaiso in Chile, a country that had been struck by disastrous earthquakes in 1731, 1822, and 1835 and would be struck again in 1960 by an 8.3 earthquake that generated a devastating tsunami. Figures 4-4 and 4-5 are pictures of some typical building damage in Valparaiso in 1906.

The buildings were disastrously designed for such a highly seismic region. Figure 4-4 shows the facade tumbling that is common when a building is not properly

Figure 4-4   Damage in Valparaiso, Chile, from the earthquake of August 16, 1906. Observe the tumbling of the front facade and the obvious lack of any cross bracing. (Photograph courtesy Branner Library, Stanford University.)

braced. Figure 4-5 illustrates the kind of damage suffered by structures that have veneers such as stucco or plaster overlaying unreinforced mud or adobe block. Such architectural ornamentation facilitates the collapse of non–load-bearing walls. A relatively simple measure like strengthening the connections at the corners of the walls would have improved these buildings' resistance to collapse.

Turkey has experienced many large and damaging earthquakes along the Anatolian fault, which extends 1000 km or so in the northern part of the country, roughly parallel to the longitudinal direction of the Black Sea. On July 22, 1967, there was a magnitude 6.8 earthquake in Mudurnu Valley approximately 200 km east-southeast of the capital city of Istanbul. Building damage was severe, particularly in multilevel structures that were not properly tied between floor levels (Figures 4-6 and 4-7). Multiple-story concrete slab buildings with inadequate reinforcements between the walls and the floors often collapse like a stack of pancakes even in a moderate earthquake. Another problem was that Turkish builders' desire to provide more open space for retailers led to a good deal of *soft-story construction*—the omission of cross-walls or frame strengthening on the ground floors of structures. When subjected to

Figure 4-5 Typical damage to an unreinforced adobe structure covered with white plaster veneer, Valpairaiso, Chile, 1906. (Photograph courtesy Branner Library, Stanford University.)

Figure 4-6   Building damage from Mudurnu Valley, Turkey, earthquake ($M =$ 6.8) of July 22, 1967. The floors and roof collapsed because of inadequate ties. (Photograph courtesy Bob Wallace, U.S. Geological Survey.)

strong ground movements, this type of building often fails disastrously at the ground-floor level.

On February 29, 1960, a magnitude 5.9 earthquake struck Agadir, an Atlantic coast port and tourist resort in Morocco. This was a moderate earthquake by seismological standards; it released 4000 times less energy than the 1906 San Francisco earthquake. Still there were 12,000 fatalities and the city suffered a surprising amount of damage. On the basis of the building damage, the relatively high intensity of X was assigned to the epicentral zone, although relatively few surface effects, such as ground fracturing or fissuring, were observed. One of the most publicized structural failures was that of the fashionable and luxurious Saada Hotel (Figure 4-8) which collapsed completely in pancake fashion, with floor slabs piled on one another (Figure 4-9). No resistance to lateral motion had been incorporated into the building's design and there was no continuity between beam-column joints, so the entire structure failed. Concrete frame buildings are usually more resistant to collapse during tremors than unreinforced masonry buildings, but when they do fail, they are more lethal.

Earthquake-resistant construction requires **bracing**—special techniques applied

Figure 4-7   Damage in Mudurnu Valley, Turkey, from the earthquake on the Anatolian fault on July 22, 1967.   Observe the collapsed-pancake result. (Photograph courtesy Bob Wallace, U.S. Geological Survey.)

to a building to prevent lateral swaying and failure during a tremor.   Three well-tested bracing techniques are frame-action, shear-wall, and diagonal or "X" bracing (Figure 4-10).

Frame-action bracing consists of installing box-shaped frames of steel or reinforced concrete in the structural framework of the walls of a building.   This type of construction is very effective in tall buildings, provided the frame is neither too rigid nor too flexible.   If the structural framework is too rigid the building will be displaced off of its foundation during an earthquake.   If, on the other hand, it is too flexible there will be excessive drift or upper horizontal displacement relative to the vertical, leading to bent elevator shafts, falling ceilings, and pounding or slamming against adjacent buildings.

All buildings are susceptible to lateral swaying.   To prevent ruinous swaying leading to collapse in an earthquake, a useful bracing technique, particularly in wood-frame buildings, is *shear-wall bracing*.   Shear-wall bracing consists of installing

Figure 4-8   View from the northeast of the Saada Hotel in Morocco prior to the earthquake of February 29, 1960. (Photograph courtesy American Iron and Steel Institute, 1962.)

Figure 4-9   View from the southwest of the Saada Hotel in Morocco after the earthquake of February 29, 1960. (Photograph courtesy U.S. Geological Survey.)

Frame action

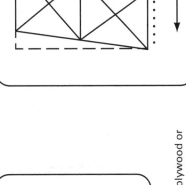

Normal position

Box-shaped frames of steel or reinforced concrete

- Effective in tall buildings
- If frame too flexible

Excessive drift

Bent elevator shafts → Falling ceilings

Pounding of adjacent buildings

Shear wall

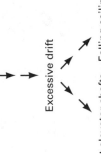

- Solid continuous walls of plywood or reinforced poured concrete to tie together vertical frame of building

  o Resists lateral swaying

  o Supports additional weight of roofs and floors

  o Good for wood-frame buildings

Diagonal or "X" bracing

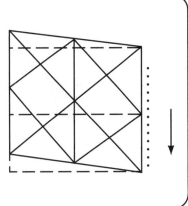

- Long sturdy strips of lumber or straps of steel attached at 45° angle across the studs of the building

- Lateral forces distributed by tension and compression in columns, beams, and cross-bracing

- Frame sections not too large

Figure 4-10 Earthquake-resistant structural forms: frame-action bracing, shear-wall structures, and cross-braced frames.

71

sheets of plywood to tie together the vertical framing of the building. The solid, continuous walls of plywood not only resist lateral swaying but also support the floors and the roof.

The third well-tested form of bracing is *diagonal* or *X* bracing. This is actually an enhanced kind of frame-action bracing in which long sturdy strips of lumber, straps of steel, or steel members are attached at 45° angles across the box-shaped frames of the structure.

4-4 DESTRUCTION BY FIRES

As all Californians know, the fires ignited by an earthquake can cause far more destruction than the earthquake itself. Even a moderate amount of ground shaking can overturn stoves, sever fuel lines, and break gas connection points, producing a large number of secondary fires. These fires can overwhelm fire departments, particularly if water pipes have ruptured and fire stations have been damaged.

On April 18, 1906, the famous San Francisco earthquake struck. The *Call-*

Figure 4-11   Refugees in Lafayette Square watching their homes burn after the 1906 San Francisco earthquake. View is looking east toward Nob Hill, a famous section of the city near the Fairmont Hotel. (Photograph courtesy U.S. Geological Survey.)

Chronicle Examiner of April 19 vividly described the conflagration that soon followed in that highly flammable city:

> Death and destruction have been the fate of San Francisco. Shaken at 5:13 o'clock yesterday morning, the shock lasting 48 seconds, and scourged by flames that ranged diametrically in all directions, the city is a mass of smoldering ruins . . . this completed the destruction of the entire district known as the "south of Market Street." . . . After darkness, thousands of the homeless were making their way with their blankets and scant provisions to Golden Gate Park and the beach to find shelter. . . . Everybody in San Francisco is prepared to leave the city, for the belief is firm that San Francisco will be destroyed. In downtown everything is in ruin. Not a business house stands. Theaters are crumbled into heaps. . . . It is estimated that the loss in San Francisco will reach from $150,000,000 to $200,000,000. [These figures are in 1906 dollars.] On every side there was death and suffering.

The city's water mains had been shattered by the earthquake, so the fire department was helpless (Figure 4-11). After it destroyed about 500 blocks of the downtown district (~ 12 km^2), the fire was finally halted when the military was called in to dynamite structures in its path. Between the tremors, the fire, and the dynamiting, there was so much destruction that 300,000 people were left without a roof over their heads.

Figure 4-12 is a panoramic view of the ruined city. Probably about 1500 people died, most of them occupants of rigid brick or stone buildings that disintegrated during the tremors. Those who lived in more flexible wooden structures survived the ground shaking, but then saw their homes burn like matchsticks in the subsequent fire.

Unfortunately, some of the lessons learned from the 1906 earthquake did not

Figure 4-12 A panoramic view of the devastation that occurred in San Francisco, 1906, as a result of the earthquake, the subsequent fire, and the dynamiting of buildings. (Photograph courtesy U.S. Geological Survey.)

register very strongly. When the Loma Prieta earthquake struck on October 17, 1989, it triggered fires in the Marina district of San Francisco—which had been built on the rubble from the 1906 earthquake. The fires could not be extinguished because of the lack of water pressure and a proper reliable water distribution system. Many older parts of the city again proved vulnerable to fire.

REVIEW

1. What is the drift of a building?
2. What are the key principles of earthquake-resistant construction?
3. In the Romanian earthquake of March 4, 1977, most collapsed buildings in Bucharest were 10- to 12-story concrete frame buildings on block corners. Why do you think corner buildings are more likely to collapse than midblock buildings? (Do not assume that corner buildings are any more asymmetric or irregular than midblock buildings.)
4. What are the critical elements of a building code?
5. Why did so many buildings collapse in the 1968 Dasht-e Bayāz earthquake?
6. What is the most important objective of microregionalization?
7. What is meant by base shear coefficient?
8. A building 20 stories high is subjected to horizontal acceleration at its base of 0.07 g or 7% g. Making use of the relation for the Uniform Building Code $V_0 = ZIKCSW$, what is the minimum value of K that should be specified so that the building can tolerate stresses produced by an acceleration of 7% g?
9. What is a safe-shutdown earthquake?
10. What are the key earthquake-bracing techniques for a building?
11. What lessons were *not learned* after the 1906 San Francisco earthquake?

5

Landslides
and Land Movement

We were driving over a mountain pass on our way back from a day at the beach. As we rounded a curve in the road, we suddenly lost control of our automobile. We thought that our front axle had broken, but then realized that it was not the car but the earth that was heaving. After 15 seconds or so of violent motion, we proceeded onward with the sense of relief and triumph people experience when they come through danger. Our elation was soon squelched when we came upon a huge mound of earth completely blocking the highway. We were trapped since we could not cross the center barrier of the highway nor easily retreat backward. We abandoned our car, hoping we could recover it intact the next day, and wondered what had caused this landslide.

5-1 EFFECTS OF LANDSLIDES

A character in a novel by the Russian writer Chingiz Aitmatov expressed the bewilderment people feel toward those treacherous plunging masses called landslides:

> What is the cause of landslides, these inevitable motions, when huge hillsides and even mountains move and collapse, widely opening the hidden abyss of the earth? And people get horrified seeing what chasm is beneath their feet. The perils of landslides . . . the catastrophe ripens . . . and only a minor vibration or rain shower is enough . . . a landslide moves formidably and there is no power or force which is able to stop it.

A landslide is the fall or downward movement of a mass of earth material to a lower level. Landslides occur in most countries. They are often started by earthquakes, particularly during the rainy season when the ground is water saturated. Earthquake-triggered landslides are both spectacular and immensely destructive (Figure 5-1).

Figure 5-1   An earthquake-triggered landslide, 200 m across at its base, from the Varto, Turkey, earthquake on August 19, 1966. (Photograph courtesy Bob Wallace, U.S. Geological Survey.)

One of the difficulties in assessing the effects of landslides and land movements is that these are often lumped together with the effects of the triggering event—which may well have been an earthquake, a volcanic eruption, excessive rainfall, or a hurricane. But landslides and other forms of downslope movement by themselves can do enormous damage to property—destroying roads, homes, bridges, dams, port facilities, airports, and recreational areas—and at times cause loss of human life (Figure 5-2).

No building, regardless of its structural configuration, can withstand the effects of downslope land movement. Much landslide damage takes place in hillside housing developments where the land has been improperly graded. In a common land-grading procedure called "cut and fill," material is *cut* from the upper slope and graded to construct a horizontal surface for placing a building. The rock debris brought downslope to produce this level surface is called *fill*. If the cut is too steep, the upper mass of soil

Figure 5-2   An earthquake-induced slide in Montana from the Hebgen Lake earthquake of 1959. The slide completely blocked the gorge, creating a lake and blocking the river's flow. Twenty-eight people were buried in this massive slide. (Photograph courtesy National Geophysical Data Center, NOAA, Boulder, Colorado.)

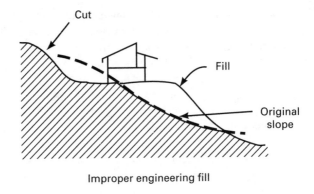

Improper engineering fill

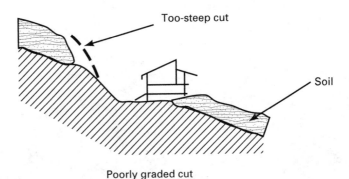

Poorly graded cut

Figure 5-3   Examples of poor hillside construction. The upper sketch illustrates the cut-and-fill procedure that builders use to make a horizontal surface. If the fill is not properly compacted, it will easily move downslope. If the cut is too steep, the upper mass of soil will easily move downhill, pushing the house off its foundation.

can move downslope, pushing a building off its foundation. If the fill is not properly compacted, it too can move downslope, destabilizing the building's foundation (Figure 5-3).

5-2 SCALES OF MOTION

Imagine a beautiful natural landscape with hills and valleys. Why does a portion of the land change and migrate downslope? What force governs downslope movement of rock and soil? The answer is *gravity*. Not the vertical component of gravity, but the component that is tangential to the potential or active slip surface. When this component exceeds the shearing resistance of the rock or soil material—either because its driving force has been increased by a steepening of the slope or because the shearing resistance of the rock or soil has been reduced—there will be downslope movement. Both causes of slope instability can be generated by natural or human agents.

Slope stability is influenced by three factors. The first is the slope's *exposure* to transient agents, such as climatic changes or earthquakes. The second can be described as the *geological factor*—the nature or makeup of the slope, its history of downward movement, and the amount of human interference it has been subjected to (has the slope been steepened by the removal of material at its base or the addition of material at its top?). The final factor influencing slope stability is the *inherent shearing resistance* of the rock material. The slope's material makeup is very important in determining the ease of downslope motion.

There are several types of landslides: rock, soil, artificial fill, and ice, rock and snow. This last type of slide is called a *snow avalanche*. An **avalanche** is a mixture of snow, ice, soils, rocks, and boulders that moves downslope at terrifying speed, annihilating everything in its path. It can produce a strong accompanying wind that tears trees from their roots. Why does an avalanche start? On a steep slope, the underlying material supporting the massive snow cover can give way because it has been soaked by spring rains or destabilized by alternating periods of precipitation and *föhn* (or *foehn*)—warm, dry winds flowing down the leeward side of the slope. Sometimes a slope is so unstable that thunder or even a loud shout is sufficient to trigger sliding of the overloaded snow (which is why skiers are asked to observe complete silence when crossing potentially unstable areas). An avalanche can travel at speeds ranging from 40 to 300 km/hr (24 to 180 mph); its velocity depends on the angle of the slope, the density of the snow mixture, and the length of its path.

Velocity is obviously an important criterion for characterizing all types of land movements. Rates of downslope speed are said to be either *rapid, intermediate*, or *slow*, with various subdivisions (Table 5-1). Rockfalls have an extremely rapid velocity range (from 3 to 100 m/sec or 6 to 200 mph). Mud and debris flows have a much wider range of downslope speeds: from a rapid 3m/sec to a very slow 6 cm/yr. The latter rate of movement is termed *creep* and such a flow is called a *creep slide*. Knowing the ground's past and present movement is useful for predicting future ground slippage. In many instances, ground distortion can be recognized prior to the onset of devastating downslope movement.

5-3 CASE HISTORIES

In a rockslide or rockfall, material moves downslope along a plane of weakness. The plane might be changed by natural deposition of sediments or weakened by human intervention. It is often thought that rockslides take place only on steep slopes. However, they can occur on slopes as small as 15° for example, a vertical drop of only 27 m (~ 81 ft) over a horizontal distance of 100 m (~ 300 ft). Though the differentiation between a rockfall and a rockslide is not precise, *rockfall* can be defined as the drop of rock from a steep cliff, while a *rockslide* takes place at lesser slope angles.

Why rockslides begin is not known, but climatic changes, particularly the

Table 5-1 Velocity Ranges for Different Types of Landslides

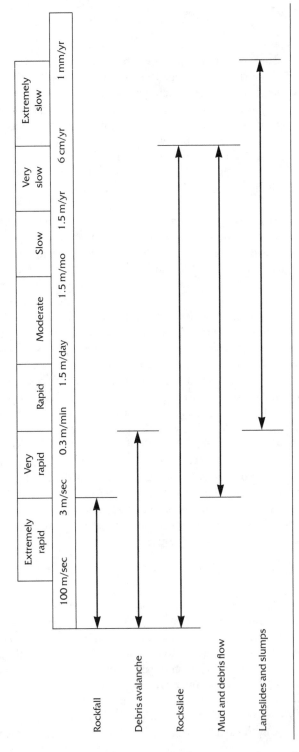

Source: Varnes (1958).

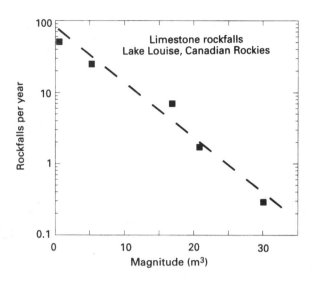

Figure 5-4 Magnitude and frequency data for rockfalls in the Lake Louise area of the Canadian Rockies. (Data from Gardner, 1970.)

freeze-thaw cycle, are assumed to be a factor. When attempts have been made to apply probability models to rockfalls it has been found that some of the observational rockfall data exhibit a magnitude-frequency correlation (Figure 5-4) analogous to those exhibited by storm surges and earthquakes. However, the data on rockfalls are limited because there have been few detailed studies covering a large time interval. Rockslides can happen without warning and have devastating consequences. Alberta, one of the western provinces of Canada, has been subjected for millennia to erosion by glaciers which has produced a steep topography susceptible to rockfalling and landsliding. When a massive rockslide struck the town of Frank, in Alberta, in 1903, about 40 million m^3 of material plunged downslope at speeds approaching 100 m/sec (200 mph) (Figure 5-5). In 2 minutes the town was obliterated.

Water plays a major role in *mud* and *debris flows*. These flows occur on slopes where heavily water-saturated soils lubricate incipient basal failure planes, decreasing their shear resistance to movement and facilitating downward flow. The rate of downslope movement is highly variable—it can be as fast as 3 m/sec or as slow as 6 cm/yr—and the material can move in surges. Mud and debris flows are common in many parts of the world, including the Czech Republic, Slovakia, Switzerland, the Philippines, and the United States. Even when they are not sudden catastrophes with very fast velocities, they can produce substantial surface deformation and damage. Figures 5-6 to 5-8 illustrate the effects of slow-moving earthflows and debris flows.

The marked, substantial decline of a mass of land along a coast or roadcut is called a *slump*. Slumping is a frequent problem where oceanfront cliffs are inter-leaved with thin layers of clay. These clay layers are easily lubricated during the rainy season, decreasing their frictional resistance to downslope slide. This is the case in the coastal regions near Santa Monica, California, where the rainy season is typically followed by slow-moving landslides or slumping (Figure 5-9). Movement is initially slow and usually can be recognized before the slump is too far advanced.

Figure 5-5 In 1903, the community of Frank in Alberta, Canada, was struck by a massive rockslide that lasted 2 minutes. In that short time, 40 million m³ of material slid downslope, burying the town. (Photograph courtesy National Geophysical Data Center.)

5-4 MITIGATION

Is it possible to reduce the hazards from landslides and other forms of downslope earth movement? One form of mitigation is personal choice: People can avoid these hazards altogether by not living in unstable areas, and if this is not feasible, they can escape personal disaster by heeding official warnings to evacuate their homes when the danger of landslide is imminent. But are there ways to mitigate or fix potential problems so that these kinds of stark personal decisions do not have to be made? Landslides are often the result of human activities, like the excessive cut-and-fill construction discussed earlier, the denudation of hillside vegetation (which facilitates downslope loss of soils), and the alteration of the pattern of natural drainage systems. One mitigative measure, therefore, is land-use restrictions.

Suppose, however, that social considerations like the need to expand habitable areas because of overcrowding make it undesirable to enact such bans in areas vulnerable to debris flows, landslides, rockfalls, and avalanches. If we cannot feasibly bar construction in these hazard-prone areas, then at least we can enforce engineering countermeasures that will make them a lot safer. Properly engineered construction

Figure 5-6 An example of a slow-moving earthflow. The horizontal ripple lines on the slope demonstrate downward movement of material. Local concentration of ground movement is evident. (Photograph courtesy National Geophysical Data Center.)

Figure 5-7   View of a debris flow that occurred on January 4, 1982, in Pacifica, California, near San Francisco. The downslope movement crushed three children to death, destroyed two houses, and damaged the house shown here. (Photograph courtesy Earl Brabb, U.S. Geological Survey.)

can slow, deflect, or trap moving mud and debris flows. The key here is proper groundwater drainage and the diversion of surface water away from gully areas, which tend to coalesce downslope drainage. For this purpose, rockfall chutes and debris runout areas can be constructed in some areas. In landslide-prone areas that are overly saturated, *electro-osmotic stabilization* can be applied. This countermeasure involves placing electrodes in the ground in positions surrounding the potential slide area. A voltage applied to the electrodes draws water from the anode (+) to the cathode ($-$) and drains the soil.

Mechanical stabilization of hillsides to prevent landslides is also possible. Revetments or barricades such as wickerwork fences and retaining walls of stone or concrete, can be constructed to stabilize downslope movements. Anchors can be inserted into firm bedrock to minimize damage from the downslope movement of more mobile overlying rock and soil material. Treatment of steep-angled slopes with

Figure 5-8   A close-up view of the debris flow shown in Figure 5-7. This debris flow was named the Oddstad Landslide because of its location on Oddstad Boulevard.   (Photograph courtesy Earl Brabb, U.S. Geological Survey.)

gunite (a mixture of cement, sand, and water sprayed onto a mold) is helpful particularly if the gunite is reinforced with wiremesh.

Studies reveal that 90% of all past landslides in the United States were geologically recognizable, predictable, and preventable. If this is so, why do we fail so often at prevention? The chief reasons that many engineering countermeasures are not implemented, especially in new-development areas, are financial and legal. What about warning systems to signal the onset of landslides and debris flows? Such geotechnical monitoring systems are in use in certain volcanic areas, but it is questionable whether every area with landslide potential can be so monitored.

There are several established defenses against snow avalanches. Wind baffles and snow fences have been installed where an avalanche might be expected to start. On downward slopes mitigative efforts concentrate either on slowing any downward movement that develops or on deflecting the path of the avalanche. Even where

Figure 5-9 Slumping in the coastal area of southern California. Note that two separate slumps can be seen. Can you recognize them? (Photograph courtesy National Geophysical Data Center.)

avalanche control is not achievable, it is possible to preserve critical transportation paths, such as highways or railroads by constructing tunnels through the base of a mountain slope subject to snow avalanches. Another measure, used on many mountains in the western United States, is to build wooden tunnels *over* railroad tracks on the downslope of potential slide areas. Finally, avalanches are routinely begun deliberately in ski areas by firing explosive missiles into the head of the slide area. This preemptive strike—carried out after the slopes have been cleared of skiers and climbers!—induces a smaller, controlled avalanche to avert a larger, more dangerous one. As a result of all of these measures fewer than 100 people worldwide are killed by snow avalanches in an average year.

REVIEW

1. What is the difference between a mudflow and a lahar?
2. Why is it difficult to determine damage done by landslides per se?

3. Why does landslide damage commonly occur in hillside housing developments?

4. What are key factors for initiating downslope movement of a mass of rock and soil?

5. What criterion would you use to assess whether future landslides in an area have the potential for becoming sudden-onset disasters?

6. Why do landslides commonly occur during the rainy season?

7. What are the main ways you can mitigate the hazards of landslides and other downslope earth movements?

8. What is the chief obstacle to developing a rapid warning system for landslides?

6

Desertification, Land Degradation, and Drought

6-1 DESERTS AND DESERTIFICATION

Every continent has a *desert*, a place where moisture is normally scant and vegetation sparse. Some deserts have river drainage—for example, the Sahara possess a skeleton of a river system—but most have interior drainage only, so that any precipitation that falls disappears into the ground or air. Most people think of deserts as regions of intense heat and endless sand, yet some deserts are cold (the Gobi Desert of China and Mongolia) and most are not composed chiefly of sand (even the famous Sahara is mostly rocky plain). Deserts are fascinating places, silent, physically challenging, often weirdly beautiful. It is no wonder that some people are drawn to them for a thrilling experience. But sometimes it is the desert that is drawn to people who have no desire to encounter it.

Desertification is a slowly advancing natural hazard rather than a sudden-onset disaster. The term is applied not only to the expansion of the sands of the desert but also to the degradation of land in the dry areas that lie along the peripheries of deserts, so that these once marginally productive soils become agriculturally useless. Desertification is a global phenomenon that has been operating for at least a thousand years.

We generally classify dry areas according to their average annual rainfall (Figure 6-1). **Hyperarid** areas receive less than 25 mm (1 in.) of rain yearly, **arid** areas get annual amounts of rainfall ranging from 25 to 200 mm (8 in.), and **semiarid** areas—the regions threatened by desertification—typically receive 200 to 500 mm

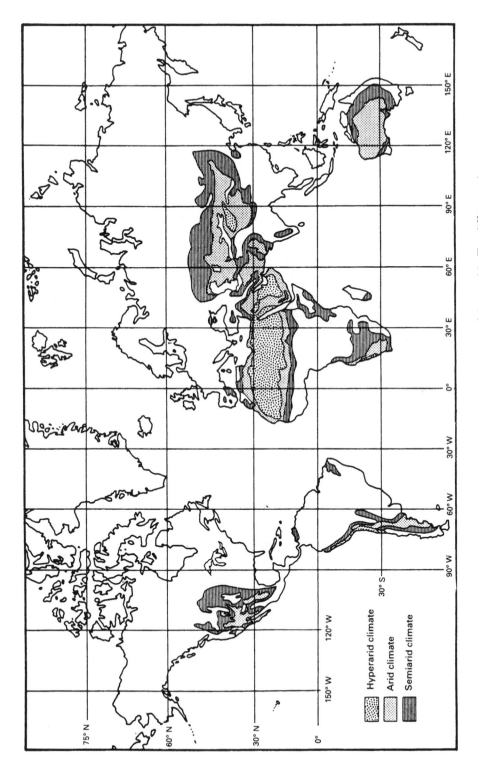

Figure 6-1 The geographic location of the dry areas of the world. The difference in annual rainfall between a hyperarid region and a semiarid region ranges between 24 mm (1 in.) and 500 mm (20 in.). (From Dregne, 1983.)

89

(20 in.) of rainfall per year. Semiarid regions are covered by a wide variety of vegetation and are suitable for some crop growing. They are particularly vulnerable to erosion, however, when their vegetative cover has been removed or degraded by careless human activity.

There are several desert regimes throughout the world. The huge Afro-Asian belt extends from the Atlantic Ocean to China. This zone includes the Saharan, Arabian, and Iranian deserts and continues eastward through Pakistan and India to the Gobi Desert in China and Mongolia. In North America the arid regions are the Great Basin and the Mojave Desert of the United States and the Sonoran and Chihauhuan deserts of Mexico. The arid lands of South America are the Atacama and Patagonian deserts of Chile and Argentina. The Australian Desert completes the global picture.

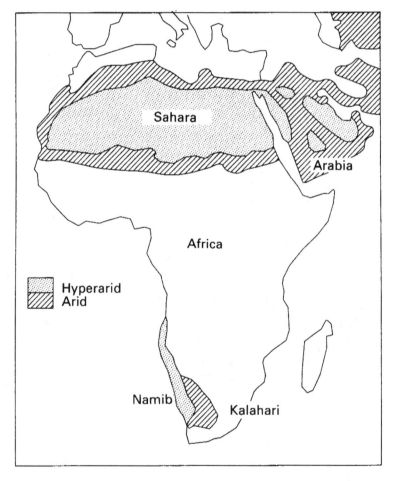

Figure 6-2 Hyperarid and arid regions of Africa and the Middle East. (Generalized from the global maps of Dregne, 1983, and Grainger, 1990.)

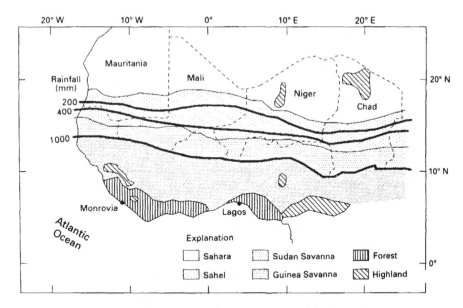

Figure 6-3   The arid zones of west Africa. (Generalized from Grove, 1978.)

Arid regions make up about 36% of the total land area of the world, or 133 million km^2 (52 million square miles); 84% of these regions are on the continents of Africa (37%), Asia (33%), and Australia (14%).

Figure 6-2 shows the distribution of hyperarid and arid climate zones in Africa and the Middle East, including the Sahara Desert and the semidesert fringe south of it known as the Sahel, the Arabian Peninsula, and the Namib and Kalahari deserts of southwest Africa. The Sahel is the region most dramatically associated with desertification today. It stretches across Gambia, Senegal, Mauritania, Mali, Burkina Faso, Niger, and Chad. Where it extends eastward into Sudan and Ethiopia, it is known as the Sudano-Sahelian region. With normal annual rainfall ranging between 200 and 400 mm, the Sahel is classified as semiarid. South of it is a wetter region of grasslands and shrubs called the Savanna. Here annual rainfall is 400 to 1000 mm (Figure 6-3). ("Wet" is a relative term. By way of comparison, a tropical rain forest receives 1800 to 4000 mm of rainfall annually.)

6-2 CAUSES OF DESERTIFICATION

Droughts by themselves do not cause desertification. Rather, it is land abuse during droughts that hastens degradation of drylands and brings on desertification. The most famous instance of desertification in the United States, the Dust Bowl of the 1930s, was created by a combination of drought and poor farming practices.

The world's drylands have very delicately balanced ecosystems that are stressed by a rigorous climate that alternates between adequate rainfall and drought. When an **ecosystem**—a functioning biophysical unit produced by the complex interrelationships between plants and animals and their environment—is irritated beyond its limits of toleration by human activity, the land becomes increasingly degraded and prone to desertification. Overcultivation, overgrazing, deforestation, and bad irrigation practices are the chief culprits in advancing land degradation. When the bad effects of these activities are compounded by climatic changes such as drought and (possibly) the greenhouse effect, the almost certain outcome is desertification.

Overcultivation is growing more crops on land than the natural fertility of the soil can support. Cultivated land is soon exhausted if soil nutrients are not replaced either by fertilizing or by natural regeneration (achieved by plowing the land under and not seeding it for one or more growing seasons). Overcultivation, leading to the loss of fertility, soil damage, and exposure to erosion generally occurs in marginal growing environments where rising populations are increasing the demand for food. It is also common where economic pressures have forced people to switch from subsistence farming of drought-resistant crops, such as sorghum and millet, to raising cash crops that require irrigation, such as ground nuts, cotton, rice, and wheat. Extensive growing of irrigated crops strains an already fragile environment.

In the Sudano-Sahelian belt *overgrazing*—too many animals cropping vegetation in the same limited areas—is a prime contributor to desertification. Uncontrolled browsing by livestock of trees and shrubs leads to soil erosion by killing the vegetation whose roots help to retain soil. Once the vegetation is gone, sand dunes from nearby deserts can migrate into the area. Also, the pounding of hooves increases fine particles in the top layer of soil, making it more susceptible to wind erosion, and compacts the secondary layers, reducing the soil's ability to drain downward. This problem is especially severe near desert waterholes.

Overgrazing is the product of ever-larger herds of animals in regions that cannot support them. Why do people keep expanding their herds in marginally productive lands? This problem is most severe in poorer countries with long traditions of nomadic life and burgeoning populations. When the caravan trade that used to sustain the nomads declined, they turned to herding to make a living. In time, they came to regard livestock as a source of wealth, which made it inevitable that they would seek to own as many animals as possible. With the expansion of populations into towns and villages, governments felt compelled to control paths of migration by placing new water boreholes along limited routes. These routes cannot sustain continually expanding herds. Forced resettlement schemes do not work well either because the villages where the herders are sent are already overcultivated and overgrazed.

Large-scale cutting of trees and clearing of brush in fragile dryland leads to massive soil erosion and desertification. This *deforestation* is a very serious environmental problem in many areas of the world today, mostly because of population pressures. Extra cropland is needed, so the ground is cleared by felling trees and burning brush. Indiscriminate grazing is another source of deforestation. Finally, trees are cut down for use as fuel. It is a frightening statistic that 50% of all the wood cut throughout the world is burnt as fuel, primarily for cooking, heating water, and general warmth. In some countries, however, wood is also an industrial fuel. Today wood furnishes 90% of the total energy consumed in the Sahelian countries. Other countries in which wood is a common fuel are India, Pakistan, and Bangladesh.

Irrigation has been practiced for thousands of years to make dry soils fertile. *Bad irrigation practices*, however, can produce desertification. The main problems are excessive irrigation and poor drainage. Either can lead to waterlogging of the soils. Then in the dry season, when the temperatures rise, excess water from the surface rapidly evaporates, causing more water to be drawn up from the substrate. Salts that were dissolved in the water remain on or close to the surface, further impeding drainage. As irrigated land becomes more and more degraded, farmers switch to salt-tolerant crops. But since the ground water is continually recharged with salty irrigation water, the land eventually becomes unproductive. The final extreme consequence is a blinding white saline desert.

Salinization has been the fate of many agricultures based on irrigation. Bad irrigation practices at least partly account for the decline of the ancient civilizations of Mesopotamia (centered around the Tigris and Euphrates rivers) and the Inca empire in Peru. Many parts of Egypt, Arabia, India, and Australia have been degraded by excessive irrigation and poor drainage. There are salt-tolerant crops that can be grown on these salinized lands, but their production is complicated by nutritional and economic questions about their worth, as well as by their lack of wide acceptability as foods.

6-3 LAND DEGRADATION IN THE UNITED STATES

The United States is not a country that comes to most people's minds when the subject of desertification is discussed, but in some regions of this country that is exactly what is going on. Remember, desertification is a complex phenomenon that encompasses more than the spread of natural deserts. Areas that are far distant from true deserts can easily degrade into barren terrain because of *poor land management* and *poor land use* in combination with uncooperative weather.

Dust storms—strong winds bearing clouds of dust across an arid region—can transport *millions of tons* of dust over large distances. The dust storms that plagued regions of the United States in the 1930s were the product of meager rainfall and years of overfarming that had destroyed the land's protective vegetation and tree cover. At times these dust-laden whirlwinds were so thick that they not only destroyed crops and killed livestock but also threatened human life. Parts of the Southwest, particularly Oklahoma, became known as the Dust Bowl. John Steinbeck, in *The Grapes of Wrath*, starkly etched these times in the story of the Oklahoma family, the Joads, who migrated, half starving, to the fertile promised land of California.

Ironically, the promised land of the Joad family is today the scene of some of the most dramatic land degradation in the United States. The San Joaquin Valley,

Figure 6-4 Salinization of a cultivated field in the San Joaquin Valley, California. (Photograph courtesy H. Wilshire, U.S. Geological Survey.)

California's agricultural heartland, is suffering from salinization for the same reason lands in the Third World are: bad irrigation practices. The use of impure irrigation waters and improper drainage will reduce the percolation rate of ground soils even in a well-cultivated field (Figure 6-4).

U.S. cattle ranchers have in some areas been as guilty of overgrazing practices as African nomads—but for profit motives rather than subsistence. The results are the same: The plant life that binds the soil is killed off and the soil blows away in the wind. Figure 6-5 shows the kind of damage uncontrolled grazing can inflict on productive land. This wasteland of gullies in the Temblor Ranges of California was created by a semidrought in wind-eroded overgrazed rangeland.

Land misuse in the United States is not confined to farmers and ranchers. Some of the country's most striking terrains are being damaged by indiscriminate recreational use, particularly our willingness to allow uncontrolled off-road vehicular traffic. When motorcycles or four-wheel-drive vehicles are ridden hard over semiarid land, they raise great clouds of dust and form deep gouges (Figure 6-6), facilitating water erosion of the now-exposed bedrock.

Figure 6-5   View of the southern Temblor Ranges of California, where over-grazing has led to the formation of gullies, or channels cut into the surface strata. (Photograph courtesy H. Wilshire, U.S. Geological Survey.)

6-4 DROUGHT: THE ORDEAL OF THE SAHEL

There is no uniform definition of drought that applies to all regions of the world. It cannot be defined simply as the absence of rainfall over a certain period of time, for there are regions, such as west Africa and parts of India, where the norm is to receive no rainfall for long intervals. Should drought then be defined as a period of rainfall deficit that exceeds the long-term average deficit interval for a particular region? The best definition, perhaps, is one based on effects: **Drought** is an extended period of rainfall deficit that results in the curtailment of the natural growth of vegetation and organisms in a region. A drought causes extensive damage to crops; a severe drought destroys crop growing altogether.

One of the great natural disasters of our time is the Sahelian drought that began in 1968 and has so far killed 100,000 to 250,000 people and 12 million cattle, as well as wrecked the agricultural base of at least five sub-Saharan countries. As the Sahara continues its grim march southward year by year, vast regions where nomadic herds-

Figure 6-6 A single motorcycle ascending a slope in Jawbone Canyon, in the northwest Mojave Desert of California. The cloud of dust was raised by the motorcycle. Observe the deep gouges in the slope. (Photograph courtesy H. Wilshire, U.S. Geological Survey.)

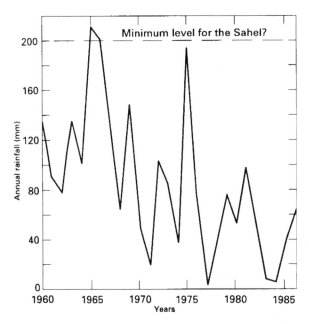

Figure 6-7 Annual rainfall at Nouakchott, Mauritania, from 1960 to 1986. The minimum annual rainfall needed for the Sahel is believed to be about 200 mm. (From Grainger, 1990.)

men and farmers were once able to scratch out a living from a few months' meager rainfall are becoming uninhabitable.

The Sahel is the southern "shore" of the Sahara Desert (*Sahel* is Arabic for "shore"). It is a transitional zone between a very arid desert to the north and forested regions to the south. Rains come into the Sahel in summer on moist oceanic air from the south, but these rains have become scanter and less reliable since 1968. Several times in this century there have been severe droughts in this region, but none lasting a quarter of a century. We can gain some idea of what has been happening by looking at a graph of rainfall in one country of the Sahel, Mauritania. Figure 6-7 shows the annual rainfall from 1960 to 1986 at Nouakchott, the capital of Mauritania. Rainfall was marginally acceptable in 1965, declined to near-nothing in the terrible years of the late 1960s and early 1970s, then returned to near-normal in 1976. It was assumed that the drought was over, but it turned out this was only a respite. As the graph shows, the declining trend in total rainfall resumed the next year, and the drought has shown no sign of abating since then.

Two hypotheses have been offered to explain the Sahelian drought: *climatic changes on a global scale* and *biogeophysical feedback*. According to one version of the first theory, the global pattern of atmospheric circulation (discussed in Chapter 7) has been altered in recent decades by increases in atmospheric dust resulting from industrial pollution and volcanic eruptions. As this atmospheric dust has cooled the middle and higher latitudes of the earth, it has distorted normal global atmospheric circulation by blocking the northward penetration of the miniature monsoons that ordinarily would bring rains to the sub-Saharan countries. Another version of the global climatic change theory proposes that the mean seawater temperature in the tropical

Atlantic has cooled, which has decreased the amount of moisture lifted from the ocean surface by the southwest monsoon winds and hence the amount of annual rainfall that falls in west Africa.

According to the second theory, drought is due to biogeophysical feedback. As the soils of the Sahel were degraded by damaging agricultural and grazing practices, the concomitant loss of vegetation led to an increase in the proportion of solar radiation reflected by the land surface back into space. As a result, the land surface cooled so that air over the Sahel constantly rises, inhibiting rainfall. If this theory is correct, the Sahelian drought is self-sustaining.

The two hypotheses we just discussed in relation to the Sahelian drought have also been used to explain drought conditions elsewhere in the world. One climatic change linked to drought in recent years is the *El Niño* phenomenon, an irregularly occurring flow of unusually warm surface water along the Pacific coast of South America. This strong current appears about twice a decade, often around Christmas (hence its name, which is Spanish for "the Infant" or Christ Child). Because it prevents the upwelling of cold deep water, El Niño raises the ocean's surface temperature and thus brings abnormally high rainfall to some usually arid areas, like the Atacama Desert of western South America. But this same phenomenon is thought to cause drought on the western side of the southern Pacific. The connection is complicated and we will only sketch it out here to give some idea of how atmospheric effects in one area may be able to produce drought on the opposite side of the globe.

The El Niño phenomenon, it has been discovered, is only part of a much larger process that affects the *entire* southern Pacific. This vast process, known as the *Southern Oscillation*, is essentially a seesawing effect between the eastern and western sides of the equatorial Pacific: When sea-level atmospheric pressure is higher than average in the east, it is lower in the west (and vice versa). This phenomenon affects temperatures on the ocean surface as well as winds and even sea level. Years of low pressure in the eastern equatorial Pacific coincide with the occurrence of El Niño. In these years ordinarily rainy Indonesia and northern Australia can experience drought conditions. The cause-effect relationship is not firmly established, however; some droughts do not correlate at all with the occurrence of El Niño.

The biogeophysical feedback theory applies to other parts of the world besides the Sahel. Poor land use, as we have seen, can alter the delicate ecological balance in semiarid environments, leading to drier land surfaces, inhibited precipitation, and the formation of dust storms.

Drought is an economic blow in some places, a vast human disaster in others. In a large wealthy country like the United States, a regional drought will damage agriculture, raise the prices of some foods for consumers, and force some farming families off the land if it lasts long enough. But no one will starve because of it—at least not today. In some African countries, by contrast, drought means severe food shortages and even famine, with massive starvation, cruel migrations, the collapse of governments, and civil wars. In the southern Sudan, Ethiopia, and Somalia, nature's fury has been matched and surpassed by human fury vented in intertribal warfare. Large tracts of land and herds of cattle have been deliberately destroyed by one side to force the other side into starvation and submission. Under these conditions the land cannot return to productivity. Instead, there is a worsening spiral of drought, land degradation, war, desertification, famine, and death.

6-5 MITIGATION

It is a matter of intense debate whether desertification is permanent or reversible. Especially where the cause is poor land management and use, it is probably not totally reversible within a reasonable time scale. It may be possible to halt it, however, but not unless we abandon the idea of an instant fix and accept the fact that only through a thorough understanding of arid and semiarid environments can we use them without ruining them.

We do know several palliative steps that can be taken: improve production by careful soil conservation and rejuvenation; substitute other fuels for wood and replant denuded lands with trees that provide windbreaks and stabilize sand dunes; discourage cattle herding. Unfortunately, these measures are hard to implement in poorer countries, where slender resources, demographic pressures, and civil wars make the people's situation desperate. The richer societies can—and do—supply emergency relief when masses of people are starving, but after the crisis is over, there is a reversion to the same practices that brought on the starvation in the first place and the cycle begins again. The only feasible solutions seem to be minimization of further land degradation, conservation of available land, and regeneration where possible. Rather than seek uncertain foreign aid for extensive irrigation and relocation schemes, these societies would do better to undertake *sustainable*, development of rural areas. This would mean favoring pastoralism over agriculture in some regions—but with animals (like camels) that adapt better than cattle to harsh, arid environments. Deforestation should be halted at all costs because clearing the land for grazing leads only to mediocre gains in the short run, followed by catastrophe.

REVIEW

1. What is the meaning of desertification?
2. Distinguish between a hyperarid region and a semiarid region.
3. What continent possesses the largest percentage of the dry regions of the world?
4. What are the primary causes of desertification?
5. Name the major types of land abuse.
6. What is a drought?
7. Identify the principal measures that can be taken to halt, and possibly reverse, desertification.
8. How are off-road vehicles contributing to desertification?

7

Atmospheric Hazards

The earth's atmosphere and oceans are interconnected fluid layers of the planet's outer shell. It is their interaction that creates the basic elements of weather, the dynamics of atmospheric motion, and two of the worst types of natural disasters: tropical cyclones and tornadoes. To understand how these atmospheric effects are produced we must know something of the global pattern of atmospheric circulation.

7-1 THE GLOBAL PATTERN
OF ATMOSPHERIC CIRCULATION

Atmospheric circulation—the movement of air masses above the earth's surface—is primarily driven by the sun. To simplify our description of the pattern of winds, we will assume that the earth is nonrotating and completely covered by oceans (Figure 7-1). Air is heated over the equator—that great imaginary circle that circumscribes the earth equidistant from the two poles. The equator is the dividing line between the northern and southern hemispheres and the reckoning datum of latitudes (the higher the latitude, the farther it is from the equator). Solar heating in the low latitudes warms the air, thereby lowering its density, so that it rises. As it does so, moistened by the high evaporation rates at the equator, it expands, cools, and condenses, producing rain. The cool (now dry) air spreads out north and south to the polar regions, where it cools further, becoming more dense in the process. Eventually it sinks to the earth's surface and flows back to the equatorial regions, where the cycle begins again. The

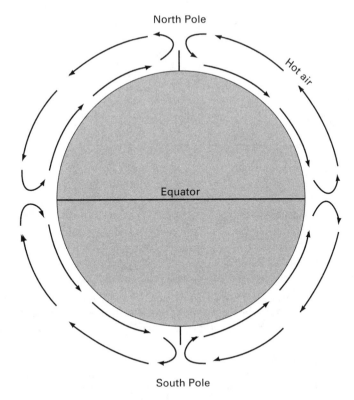

Figure 7-1   On a nonrotating water-covered earth, air would rise at the equator and be cooled at the poles. It would then sink and return to the equator.

prevailing winds are toward the equator in both hemispheres. Thus in the northern hemisphere the winds blow from north to south, and in the southern hemisphere the direction is reversed. Such a simple, monotonous pattern will produce no storms, no sultry summers or frigid winters.

Actual atmospheric circulation is far more complicated because of the existence of the continents, the rotation of the earth around the sun, and the tilt of the earth's axis of rotation. The earth's *axis of rotation* tilts $23\frac{1}{2}°$ from the plane of its elliptical orbit about the sun (Figure 7-2). Because of this tilt or inclination, the vertical rays of the sun attack $23\frac{1}{2}°$ north latitude (the *Tropic of Cancer*) in late June; this is known as the **summer solstice**, the time of the year when the sun's vertical rays are at their most extreme northward reach. The vertical rays of the sun attack $23\frac{1}{2}°$ south latitude (the *Tropic of Capricorn*) in late December; this is the **winter solstice**, the time of the year when the sun's vertical rays are at their most extreme southward reach. In the northern hemisphere the most daylight hours occur at the summer solstice and the least daylight hours at the winter solstice. In the southern hemisphere the situation is reversed.

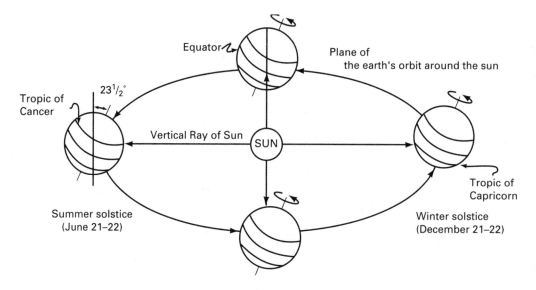

Figure 7-2   The earth's orbit around the sun. Because the earth's axis tilts $23\frac{1}{2}°$ from the plane of its orbit around the sun, vertical rays of the sun attack at different latitudes during a yearly cycle. (From Lutgens and Tarbuck, 1992.)

Thus the tilting of the earth's axis of rotation produces differential heating of the earth's surface over the course of the year, and hence the seasonal changes.

The earth's *rotation* complicates the pattern of atmospheric circulation through the **Coriolis effect** (named after the nineteenth-century French mathematician who discovered it). To an observer on the earth, an object moving above the surface of the earth does not proceed in a straight line but is deflected—to the right in the northern hemisphere, to the left in the southern hemisphere. Because the earth's rotation is counterclockwise, every object moving on or above the earth moves to the east at a velocity that is dependent on latitude. This velocity is greatest at the equator, where the circumference of the earth is greatest, and decreases progressively at higher latitudes until, at the poles, it is zero, since an object there undergoes no horizontal displacement on the rotating earth.

To visualize the *Coriolis effect,* picture a projectile being launched from the equator toward the North Pole (Figure 7-3). Because of the counterclockwise rotation of the earth, the projectile—*before it is fired*—possesses an eastward component of velocity of ~ 1700 km/hr (1000 mph). After the projectile is fired, it will still have this component of velocity, together with the speed given it at its launching. But as the projectile moves northward it is moving over the regions that are rotating eastward at progressively slower speeds than at the equator. At 30° N latitude, for example, the eastward component of velocity of the earth's surface is ~ 1500 km/hr (~ 900 mph). But the projectile has an eastward component of velocity of ~ 1700 km/hr, which it acquired at the equator. Thus for someone standing on the earth and observing the

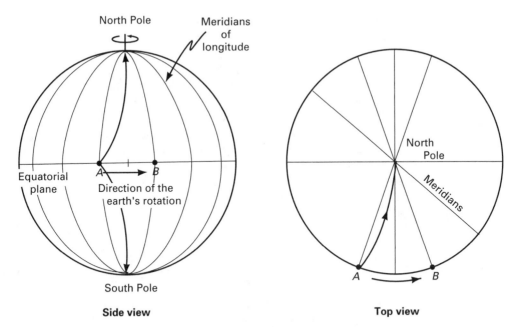

Side view   **Top view**

Figure 7-3   The Coriolis effect: During a rocket's flight from the equator at point *A* to the poles, point *A* appears to have moved right to the position denoted by *B*. This is because the rotation of the earth produces a curved path for the rocket's path when plotted on the earth's surface. The situation is reversed in the southern hemisphere: There the apparent movement would be to the left.

flight path, it would appear that the projectile path is being deflected to the right. Someone stationed on the moon looking down on the earth, however, would see a projectile moving in a straight-line trajectory with the earth rotating beneath it. Were the projectile launched from the equator toward the South Pole, the deflection would also be to the east, but to an observer on the earth, it would appear to be to the left because in the southern hemisphere the earth appears to have a clockwise rotation. If the projectile were fired parallel to the equator, of course, there would be no Coriolis effect.

A more mundane but amusing result of the Coriolis effect is that water drains from a bathtub in different directions depending on what hemisphere you are in. After watching it spiral and drain counterclockwise all my life, I could not wait to observe on a trip to Rio de Janeiro that water indeed drains clockwise in the southern hemisphere.

Wind blows because of differences in atmospheric pressure between areas. How does the Coriolis effect influence the circulation pattern of the earth's atmosphere and hence its winds? Recall that in our idealized simple picture of atmospheric circulation air masses move from the equatorial regions to the poles and back again. In actuality, moving air is deflected by the Coriolis effect, just as any other object moving on or above the surface of the earth is. Winds flowing *toward* the equator are deflected west (to the apparent right in the northern hemisphere and to the apparent left in the southern hemisphere), while those blowing *away* from the equator are deflected east. In both cases, the amount of deflection increases toward the poles. This sets the prevailing east-west direction of winds on the earth's surface (Figure 7-4). Warm air moving northward and southward from the equator produces a band of wind in each hemisphere that is coming from the west at high altitudes in the mid-latitude range. These winds, called *westerlies,* are at a latitude of 30° N and 30° S. To the north, they meet the *easterlies* coming from the polar highs. To the south, the poleward-flowing air—dry because its moisture was released near the equator—sinks toward the surface of the earth in subtropical high-pressure belts where the winds are light and variable, the weather normally calm and cloud-free. These high-pressure zones in both hemispheres are called the *horse latitudes*, possibly because old-time sailors believed the long delays in these notorious calms were fatal to horses being

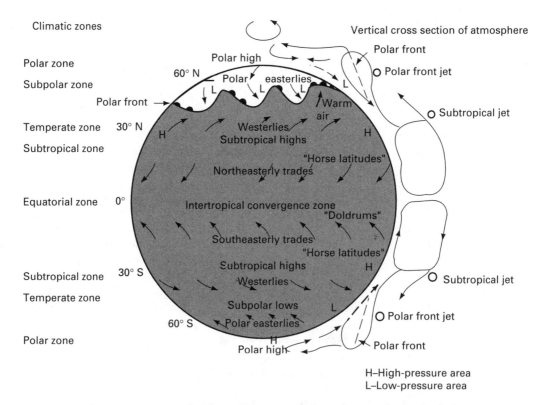

Figure 7-4 A schematic representation of atmospheric circulation on a rotating earth. The circulation cells shown are the two thermally driven circulation systems on each side of the equator. The returning flow to the equator is deflected by the Coriolis effect, producing the trade winds. (From Gross, 1987.)

transported across the ocean. (An alternative explanation of the name's origin is that on ships becalmed in the blistering heat of these zones sailors were inclined to throw cargo horses overboard to conserve water.)

The surface winds that flow almost continually from the horse latitudes toward the equator are deflected toward the west in both hemispheres because of the Coriolis effect. These are the famous **trade winds** known to Columbus and other early mariners for their reliability. Called the *northeasterly trades* in the northern hemisphere and the *southeasterly trades* in the southern hemisphere, they occupy most of the tropics and are a major component of the global pattern of atmospheric circulation. The intertropical convergence zone where these two sets of trade winds converge is known as the **doldrums**. This equatorial belt is characterized by light, variable winds and deep calms broken by sudden heavy rains or squalls.

Finally, a word about the effects on atmospheric circulation from land masses— also left out of our original simplified description. The *presence of continents* modifies atmospheric circulation by establishing seasonal patterns of high- and low-pressure centers. Because water changes temperature far more slowly than land, land masses are colder than oceans in winter and warmer in summer. As warm air rises in the lower-pressure areas over continents in summer, it attracts air from the higher-pressure areas offshore, resulting in the pleasant sea breezes enjoyed in many coastal areas and the wet, stormy monsoons of Asia. In winter these pressure centers are reversed: Cold dense air sinking over land masses creates high-pressure centers so that air is drawn away from the continents toward the now-lower-pressure centers over the oceans. One important effect of this is the dry winter monsoons of Asia.

The pattern of atmospheric circulation has a profound effect on the formation of tropical cyclones and the weather conditions that produce tornadoes and other disturbing atmospheric effects.

7-2 TROPICAL CYCLONES

Tropical cyclones (called **hurricanes** in the western Atlantic and Caribbean and **typhoons** in the western Pacific) are whirling masses of wind and rain wrapped around a low-pressure center. These great storms spawned in the tropical seas move with furious speed and destructive force, producing much human suffering and devastation on islands and along the coastal areas of continents. Figure 7-5 is a global picture of areas of tropical cyclone formation. Note they are generally confined to latitudes 15° to 20° north and south of the equator.

Lands to the west of oceans are most at risk. India and Bangladesh are vulnerable to Indian Ocean cyclones, the Philippines, Japan, China, and Australia to Pacific typhoons, and the Caribbean islands, the southeastern United States, and Mexico to hurricanes roaring out of the south Atlantic. When Hurricane Gilbert swept northwestward from the tropical Atlantic Ocean into the Caribbean Sea and the Gulf of Mexico in September 1988, it devastated the island of Jamaica, raged through the resorts of Cozumel and Cancún on the Yucatán Peninsula of Mexico at 280 km/hr (170

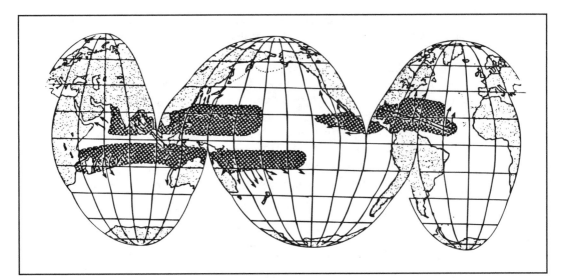

Figure 7-5   Areas of tropical cyclone formation and paths of some past cyclones. Tropical cyclones originate in the belt of trade winds that move the storms from east to west and then toward the poles, where they are deflected by the westerlies. (From Gross, 1987.)

mph), and then turned to strike the northeast coast of Mexico, south of Brownsville, Texas. By the time the hurricane was spent, 315 people were dead and tens of thousands were homeless.

More savage still are the Pacific typhoons that regularly batter the Philippines and Japan. These are often larger, and always more numerous, than Atlantic hurricanes. In an average year several typhoons strike one or more of the major islands of the Philippines. In the unaverage year of 1970 the Philippines were swept hard by four typhoons that killed 1500 people. Japan is hit by about four typhoons a year, which regularly cause millions of dollars' worth of damage. The ferocious typhoon named Vera that assaulted central Japan in September 1959 left 5000 dead and eight times that many injured, and crippled the Japanese railway system as well.

The deadliest of all tropical cyclones are those that originate in the Indian Ocean and ravage the thickly populated Bay of Bengal coasts of India and Bangladesh. Because of its funnel-like shape, the Bay of Bengal relentlessly channels storm waves into Calcutta and other coastal cities in northern India and the low-lying country of Bangladesh. In 1970 a violent typhoon slammed into Bangladesh (then East Pakistan), lifting the sea over 15 m in places and causing a catastrophic inundation in which at least 300,000, and perhaps 1 million, people perished. Countless flimsy dwellings were smashed to bits and thousands of tons of rice washed away in this the most crowded and one of the poorest countries on earth.

How do tropical cyclones begin? The process starts over very warm water near the equator. Heated air is carried aloft to the higher reaches of the atmosphere, where it creates a tropical depression—a region of low pressure that pulls in colder, drier air from the surrounding atmosphere. Subject to the Coriolis effect, these inward-flowing or centripetal winds do not move in a straight line but spin counterclockwise (in the northern hemisphere) or clockwise (in the southern hemisphere). Note that a tropical cyclone cannot form *too* close to the equator because the Coriolis effect does not operate there.

As the moist air condenses into spiraling clouds and cools, two things happen. First, it cannot hold as much water vapor as it did when it was warmer, so rain begins to fall. Second, the condensation releases heat, increasing the amount of humid air in the region above the storm system, which further reduces pressure near the center or eye of the storm (Figure 7-6). As more surface air is sucked into the vortex, it can only travel upward, repeating the cycle of creating faster winds, bringing in more moist air, and producing further rainstorms. It is the pressure gradient that generates the rapid, inward-spiraling winds of a tropical cyclone. As the air moves into the center of the storm, its rotational velocity increases because of the conservation of angular momentum. This principle is easy to grasp by visualizing a figure skater rotating with both arms extended. As the skater draws his or her arms inward for a dramatic finale, the rate of spinning rapidly increases. Just so, a tropical cyclone is at first a tightening ring of storm clouds with extensions of curved cloudy arms. As the "arms" are pulled in by suction supplied by winds in the upper atmosphere, the cyclone gathers maximum speed and force. Meteorologists have created several scales for indicating wind

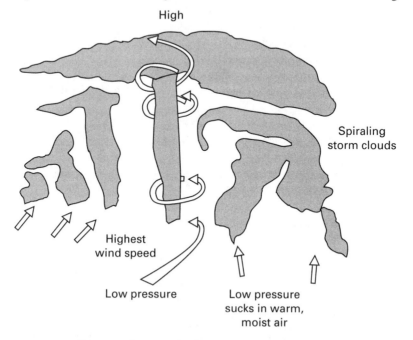

Figure 7-6   Diagram showing the formation of a tropical cyclone or hurricane.

speeds. On the earliest such scale, the Beaufort scale of ocean wind speeds (Table 7-1), a storm is designated a hurricane when its wind speed exceeds 120 km/hr (~ 75 mph). The National Hurricane Center (Table 7-2) classifies hurricanes into five categories depending on wind speeds, which have been recorded as high as 250 km/hr (~ 155 mph).

At the core of a tropical cyclone is its eye—a roughly circular area of light winds and fair weather several miles in diameter. This strangely tranquil core walled by spiraling violent winds normally migrates northwestward, though some cyclones make baffling changes of direction, even doubling back on themselves. Most storm systems in the northern hemisphere, however, travel westward and poleward. As the cyclone moves north from the equator, the Coriolis effect enhances the angular momentum of the winds circling the eye, feeding the storm as it moves. The rate of movement of the eye ranges from 24 to 80 km/hr (15 to 50 mph), and the cyclone can travel long distances in the open ocean before it dissipates.

Once a tropical cyclone travels outside the zone of hot tropical seas, it weakens, either because it strikes land where there is dry air or because it approaches cooler ocean surface temperatures that rob it of its supply of warm, moist air. Even in its weakened state, the storm's winds can be severe and it can release huge amounts of rain. The winds do great damage, but it is the *storm surges* of a tropical cyclone that are most devastating. As the low-pressure eye of the storm approaches a shoreline, the mean water level rises because of the decrease in atmospheric pressure. When the storm passes, the water level returns to normal, producing a landward wall of water—or storm surge—that sweeps miles inland. Storm surges are responsible for the large majority of tropical cyclone fatalities. In the worst possible scenario the storm surge hits the coast at high tide. This is what happened in the Galveston, Texas, hurricane of 1900, probably the worst in U.S. history: The waters of the Gulf of Mexico at the Galveston Island shoreline rose 4 ft in *4 seconds*. The estimated number of fatalities was 6000.

Offshore or barrier islands like Galveston and low-lying coastal areas like those of Bangladesh are most vulnerable to storm surges. Much of Bangladesh's coast is

Table 7-1   The Beaufort Scale of Ocean Wind Speeds

Rank	Wind description	Wind speed (mph)
0	Calm	< 2
1–3	Light breeze	2–12
4–5	Moderate wind	13–23
6–7	Strong wind	24–37
8–9	Gale	38–55
10–11	Storm	56–75
12	Hurricane	> 75

NOTE: This scale was devised by Captain (later Admiral) Sir Francis Beaufort of the British Navy in 1805. The original table, invented to record operating conditions on the high seas in ships logs, made no reference to actual speeds. These were added in 1906, and there have been several modifications and additions since then. This is the 1944 version. (1 mph ~ 1.6 km/hr.)

Table 7-2 The Ranking of Hurricanes in North America and the Caribbean by the National Hurricane Center According to Wind Speed

	Category 1	Category 2	Category 3	Category 4	Category 5
Barometric pressure (in.)[a]	>28.94	28.50–28.91	27.91–28.47	27.17–27.88	<27.17
Wind speed (mph)[b]	74–95	96–110	111–130	131–155	>155
Storm surge (ft)[c]	4–5	6–8	9–12	13–18	>18

[a]Unfortunately, the units used to measure barometric pressure are not uniform across the globe. U.S. usage favors inches of mercury observed in a barometer. At sea level, normal barometric pressure is 29.92 in. or 760 mm of mercury. Meterologists equate ~ 750 mm of mercury with 1000 millibars or 1013.6 hectopascals.

[b]1 mph ~ 1.6 km/hr.

[c]1 ft ~ 0.3 m.

less than 2 m above mean sea level. When the 15 m storm surge of November 13, 1970, was added to normal high tide, the result was one of the worst natural disasters of modern times. In 1991 another tropical cyclone struck Bangladesh, claiming at least 138,000 lives. Bangladesh's bad geographic luck is compounded by poverty and overpopulation. Its people have spread into the hazardous low-lying areas because the land there is fertile and ideally suited to rice cultivation. Most of them live in fragile houses of straw and bamboo that are easily destroyed by any strong storm, let alone a tropical cyclone. The country, like most poor nations, does not have the infrastructure to establish effective evacuation procedures.

7-3 TORNADOES

Winds do not have to operate on the gigantic scale of tropical cyclones to do immense damage. Tornadoes are very much smaller than tropical cyclones—typically only about 40 m (~ 130 ft) in width compared with the 10,000 to 100,000 sq km area of a cyclone—but they are extremely violent and destructive. **Tornado** originally meant a violent squall blowing outward from the front of a thunderstorm along a portion of the west African coast, but today it usually refers to the type of extremely violent small-diameter revolving storms found east of the Rocky Mountains in the United States, particularly in Texas, Oklahoma, Kansas, and Missouri. In fact, this portion of the central United States is known as "Tornado Alley" (Figure 7-7). Tornadoes commonly rake this region in the afternoons or evenings in the months of April, May, and June. Although 80% of all tornadoes occur in the United States, other areas of the world are not immune. Minitornadoes form during winter and spring in the southeastern coastal areas of Australia; these are not as destructive as U.S. tornadoes because their wind speeds are lower. (Australians refer to their tornadoes as *willy-willys* or *cockeye bobs*.) Occasionally tornadoes strike in Europe, Africa, and on the Indian subcontinent.

A tornado looks like a **vortex**—a swirling or spinning mass of fluid—descending from a storm cloud and tapering long and narrow toward earth. The inward-and-upward whirl of the winds explains why tornadoes are popularly known as "twisters" in the United States. The lateral movement of a tornado is faster than that of a tropical cyclone, typically ranging from 50 to 200 km/hr (~ 30 to 120 mph), and internal wind speeds can be as high as 500 km/hr (300 mph)—about double the wind speed of the most powerful hurricanes.

For all their violence, tornadoes usually last only a minute or two and their destructive path rarely exceeds 10 km (6 miles). Their routes are very narrow and can be discontinuous. During the passage of a tornado, funnel clouds have been observed to recede upward and then drop down again like a hammer. Barometric pressure within the eye of a tornado can suddenly fall as low as 800 hectopascals or ~ 590 mm of mercury—that is a sudden drop of ~ 21%! Such a pressure drop creates a force that can lift automobiles and roofs skyward and cause a closed building to explode outward.

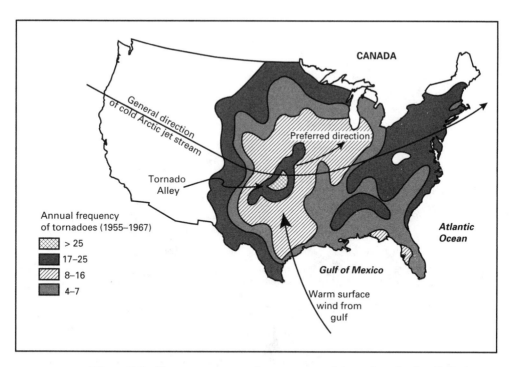

Figure 7-7 The average annual occurrence of tornadoes in the United States. Cold Arctic air flows over the Rocky Mountains to mix with warm surface air coming from the Gulf of Mexico. This causes the formation of "Tornado Alley" a path extending from Texas and Oklahoma through Kansas and into Missouri. (Generalized from Eagleman, 1983.)

In 1928 a Kansas farmer named Will Keller looked into the eye of a tornado before he closed his cellar door:

> Everything was as still as death. There was a strong gassy odor and it seemed that I could not breathe. There was a screaming, hissing sound coming from the end of the funnel . . . . I looked up and to my astonishment I saw right up into the heart of the tornado. There was a circular opening in the center of the funnel, about 50 or 100 feet in diameter, extending upward for a distance of at least one-half mile.

Why do tornadoes form? Tornadoes are always associated with thunderstorms. Two rotating air masses, one hot and humid from the tropics and the other a slowly moving cold front from the north, move toward each other. If neither gains substantial momentum on the way, there will be nothing but a thunderstorm when they collide. But if either or both of these revolving masses gather force, their encounter will be far more violent. Figure 7-8 is a schematic representation of the formation of a tornado. The influx of warm, moist air in front of a thunderstorm from the southeast opposes the higher-level, heavier cold flow from the northwest. As the cold air meets

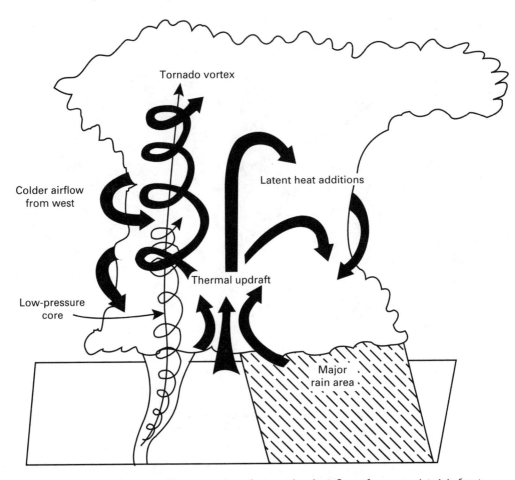

Figure 7-8   The generation of a tornado: An inflow of warm moist air in front of a thunderstorm opposes a higher-level northwesterly flow to form a cyclonic vortex. (After Eagleman, 1983.)

the rising thermal updraft of the warm mass, the air spins faster and faster. In this wild turmoil a cyclonic vortex is born.

The United States seems to have the perfect geography for hatching tornadoes. Every spring cold air moves down from the Arctic, condensing into rain clouds over the Coast Ranges of the West Coast. By the time this cold, moving air mass has crossed the Rockies into the Great Plains, it is wrung dry. The humid Gulf Coast, meanwhile, produces torrential rains along the Texas coast. The meeting of these two air masses in Tornado Alley has given rise to some memorable tornadoes. In April 1974 such an encounter generated 148 twisters in less than 24 hours, which whipped across 2600 miles and 13 states, killing over 300 people. In the southeastern United States multiple tornadoes are also set up in the southeast quadrant of a slowly moving hurricane. This kind of tornado formation occurs during late summer and autumn, and the spatial track of the tornado's eye is in an easterly-northeasterly direction (Figure 7-9).

Because its path is short and narrow, the average tornado is not responsible for a high number of deaths. Property damage also tends to be limited, for the same reason. Severe tornadoes or tornado clusters like the April 1974 phenomenon, however, have killed hundreds and destroyed whole towns. Like a giant vacuum cleaner in the skies, a large tornado can lift heavy trains, great-rooted trees, iron bridges, and concrete piers and hurl them down many feet—occasionally miles—away. Cows, horses, and people have been whisked up and carried through the air like Dorothy in *The Wizard of Oz*—with much more brutal consequences.

Before we leave the subject of tornadoes, we should say a word about the *unnat-ural* tornado phenomena brought into being by the Chernobyl nuclear accident of 1986. The monumental explosion of this Russian plant created large thermal plumes that rose up into the atmosphere, where their tremendous energy was sufficient to form tornado like vortexes in interaction with only light to moderate winds.

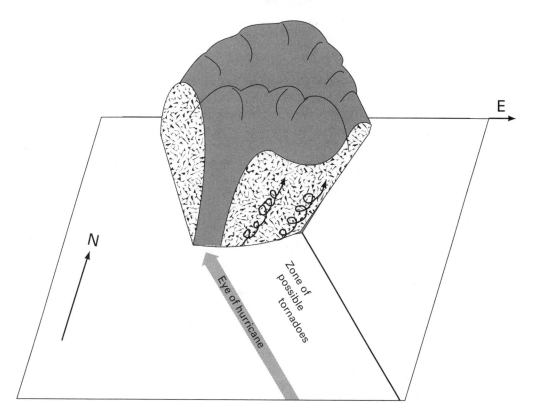

Figure 7-9   Tornado setup in the southeast quadrant of a hurricane (the lead-ing right-hand quadrant).

7-4 SMALLER-SCALE WIND PHENOMENA

Dust devils are small whirlwinds of dust and sand commonly found in arid and semi-arid regions of the world. Since surface heat is necessary to their formation, they form from the ground up. A dust devil can turn into a disturbingly strong vortex, as I learned during a reconnaissance trip into the eastern Hashemite Kingdom of Jordan to select a field site for an earthquake monitoring station. My colleagues and I saw a small cloud of rotating dust approach us. When it reached the oasis where we were standing, it enveloped and sandblasted us—leaving us completely disoriented (Figure 7-10). Since the sun was directly overhead, we could not immediately tell any of the cardinal directions.

Warm air from extreme surface heating during the summer is the main contributor to the formation of these sinister phenomena. Their vortexes range human size to tens of meters or so—too small to be governed by the Coriolis effect, so dust devils can rotate either counterclockwise or clockwise in the northern hemisphere. They are typically only a few meters in diameter with heights of ~ 100 m. Wind speeds can reach 100 km/hr (60 mph), but dust devils are short-lived, usually lasting only several minutes (it is rare for a dust devil to persist for an hour or more). This is not a fatal hazard, then, but when a dust devil appears to track you across the open plain or desert, it can be a terrifying experience.

Strong updrafts of air resembling dust devils but modified by the variation in topography are found in mountainous regions bordering plains. One notable site for

Figure 7-10 A dust devil observed in eastern Jordan. (Photograph by Robert L. Kovach.)

these *mountainados*, is the area near Boulder, Colorado, that borders the eastern front of the Rocky Mountains. Unlike dust devils, mountainados can form in winter as well as in summer, if the thermal conditions are just right.

Dust storms are wind storms without accompanying rainfall. They are known across the world, from the pampas of Argentina and southern Australia to the Sahara and Sahel of Africa, from the Russian steppes to Pakistan to the plains of Canada and the United States. We discussed dust storms in Chapter 6, in relation to droughts, and noted how a combination of low precipitation, poor land use, and improper irrigation can produce dry storms that devastate agriculture.

We conclude this section on smaller-scale wind phenomena with a note about the winds that started uncontrollable fires in southern California in the fall of 1993. These hot, dry winds, called *Santa Anas* after the canyon south of Los Angeles where early settlers first observed them, start when a clockwise rotation of a giant high-pressure system to the east and north, over the Utah desert, reverses the normal airflow from the west. In October–November 1993 the Santa Anas rushed through the mountain canyons of southern California on their way to the ocean. As the air descended and compressed, it grew extremely dry, creating tinderbox conditions in vegetation already dried out by the summer drought. This combination of hot winds gusting up to 112 km/hr (70 mph), and desiccated vegetation produced wildfires that raged through Altadena, Laguna Beach, Malibu, and other areas near Los Angeles. The Santa Anas were particularly treacherous that year because the previous very wet winter had been followed by massive growth of vegetation—which in the early autumn became volatile fuel.

7-5 MITIGATION

Most of the atmospheric hazards discussed in this chapter cannot be prevented. Dust storms are an exception to some extent—because they are often instigated by careless treatment of the soil; the soil-conservation measures noted in Chapter 6 would sharply reduce their occurrence. And the unnatural tornado like phenomena produced by Chernobyl (and other industrial disasters) can certainly be prevented by human actions since they are the products of human acts. But tropical cyclones and tornadoes are pure forces of nature—hazards we must live with. Some have proposed fighting tropical cyclones with technological fixes such as seeding the cyclone with silver iodide to disperse the clouds and weaken the storm, but nothing much has come of these modification schemes. Besides, tropical cyclones are necessary to some countries' agriculture (Japan, for instance, gets about 25% of its rainfall from typhoons). If we could prevent or modify them in one area of the globe, it might cause drought in another region.

The only real goal is to minimize the disastrous effects of tropical cyclones and tornadoes by giving timely, accurate warning to populations, by using construction techniques that ensure buildings have a good chance of surviving, and by preparing an effective disaster response.

Since warning systems are national and regional, they vary in monitoring ability from place to place. No warning system is infallible, but tornado warning systems in the United States are now fairly sophisticated, particularly in local areas. Tropical cyclone warning systems are less reliable because of the huge scale of these phenomena and the unpredictability of their movements.

It has been found time and again that well-designed *steel-reinforced* concrete and stone buildings, low in profile, generally survive tornadoes and tropical cyclones. We also know how to build excellent tornado shelters such as specially reinforced basements.

Disaster responses, too, are under our control. Hurricane Andrew had been watched closely for at least a week before it hit Florida with catastrophic force and reduced parts of the state to rubble in September 1992, so there was plenty of time to prepare a response. But efforts by both the state and the federal government were slow and disjointed. People whose homes had been flattened by the storm were without food, drinkable water, and shelter for up to a week. The Federal Emergency Management Agency (FEMA) performed so poorly that it was accused of being "brain-dead," and the state did not do much better. This and other bad performances apparently goaded FEMA into a bureaucratic overhaul, for in the Great Flood of 1993 in the Midwest, the agency swung into action much faster and more effectively. So while earth's fury is unmanageable, our preparations against it are not.

REVIEW

1. What are the three main factors controlling the global pattern of atmospheric circulation?
2. How does a tropical cyclone originate?
3. Why does a tropical cyclone weaken when it moves outside the zone of tropical seas?
4. Why do tropical storms that originate within 5° of the equator not acquire as much of a rotary motion as storms or cyclones originating at somewhat higher latitudes?
5. A tornado has higher wind speeds than a hurricane, yet a hurricane produces more total damage. Why?
6. Why does a tornado form?
7. When you receive a tornado warning should you board up your windows? Explain.
8. What are dust storms and why do they occur?
9. What is a dust devil?
10. Do tropical cyclones have any beneficial effects?

8

Oceanographic Hazards

Rhythmic and soothing one day, ferocious the next, the vast seas have always exerted a near-mystical fascination. Over two-thirds of the earth's surface is covered by oceans—a startling abundance of water compared with what we know of other planets. All civilizations, even those totally land-bound, have depended on trade carried back and forth across the seas. In fact, the earliest students of oceans were the intrepid mariners who made this trade possible. Long before the invention of modern oceanographic instruments, there was a quite sophisticated body of knowledge about ocean currents, wind patterns, and oceanographic hazards derived from observations made on such voyages.

8-1 AN INTRODUCTION TO WAVES

The main oceanographic hazards, both on the open seas and along the shore, are generated by waves. A *wave* is water subjected to pressure from a variety of phenomena—wind, seismic and volcanic disturbances, and the gravitational interaction between the earth and the moon. We are concerned here with wind-generated waves, though much of what we say is applicable to the other types of waves as well.

Ocean waves come in all sizes, from small ripples to massive swells that batter even large ships on the open seas and slam into shorelines with all the sea's weight and destructive power behind them. Wind is the main mechanism of wave production on the water surface. The size of a wind-generated wave will depend on the wind's speed, the length of time it is blowing in a particular direction, and its **fetch**, or the geographical distance over which it blows in a constant direction. Storm surges are the most hazardous type of wind-produced wave. The waves generated by the gravitational attraction between the earth and the moon give rise to the normal high and low tides observed along coastlines. Undersea seismic disturbances cause one of the most dreaded of all natural hazards: the tsunami or seismic sea wave. The technical details of wave propagation in the oceans are complicated, but the critical aspects of this movement of the water surface can be easily understood. When wind blows on still water, the surface of the water is immediately covered by a pattern of crests or undulations that travel slowly in the direction of the wind. A water wave has height, length, and speed (Figure 8-1). The height of a wave is the vertical distance between its crest (highest point) and its adjacent trough (lowest point). **Wavelength** refers to the horizontal distance between adjacent wave crests or troughs. **Wave period** is the interval of time between the passage of successive crests or troughs of a wave, while **wave frequency** (the inverse of wave period) refers to the number of times a wave's crests or troughs pass a fixed point within a specified interval of time. For waves in the deep water of the oceans, speed, height, and wavelength depend on the wind's velocity (meaning its speed *and* direction) and the water's depth.

In the open seas wave height is assessed by noting how high above the ship's waterline an observer must be to see the passing wave crests "top" the horizon. Ships at sea routinely collect this sort of information, which has always been used by mariners to chart shipping lanes and avoid high-sea regions. Today orbiting satellites record wave heights in the open ocean, so we can monitor the heights and seasonal fluctuations of waves on a global scale.

The speed, C, of an ocean wave is equal to the product of its wave frequency, F, and its wavelength, L, that is, $C = FL$. Wave speed is proportional to water depth: The deeper the water, the faster a wave can travel. A typical deepwater ocean wave travels at a speed of ~ 700 km/hr. As a wave approaches the shoreline, it slows down because of the shallowing of the water. Wavelength equals speed times wave period; that is, $L = CT$. So when the wave period is 15 minutes ($\frac{1}{4}$ hour), the wavelength of the

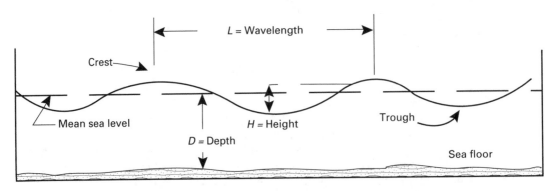

Figure 8-1   The geometry of a water wave.

wave is 175 km, and when the wave period decreases to 1 minute, the wavelength shrinks to a little less than 12 km.

As the wave's speed decreases and its wavelength is compressed, there is no reduction in the wave's energy. A wave's energy is equal to the square of its height times its wavelength, or H^2L. Since energy always remains constant in a column of water, the compression of the wavelength results in an increase in the wave's height: The energy is displaced from the horizontal (wavelength) to the vertical (crest). That is why waves that are placid out at sea rear up and hurl themselves against the shore and why larger deepwater waves can turn into hazardous monsters when they smash into land.

Other factors besides speed help determine the impact of a wave. Among these are **wave velocity**, which is the wave's speed plus the direction of its approach; the degree of shallowing of the ocean bottom; the location of promontories, tidal inlets, harbor entrances, and breakwalls; and wind velocity and turbulence. Especially important is the topographical (configuration) of the shoreline, which governs wave shoaling, refraction, and diffraction.

Wave shoaling is the movement of a wave from a greater to a lesser depth of water. As a wave moves in toward shore, it is affected by the slope of the ocean bottom. How a shallow-water wave "feels the bottom" determines how it will slow down, peak, and break—what kind of breaker it will become (*breaker* is the term used for breaking waves in shallow waters). A wave's peak or *breaking height* is its maximum height relative to the forward trough. Its *breaking depth* is determined by the distance between the sloping sea bottom and the level of the still water near the shoreline (Figure 8-2). Where the sea bottom's slope is slight, waves rise only gently and break very gradually. These *spilling breakers* spread out and dissipate their energy over a wide expanse. Where the gradient is steeper, the breaking waves rise in a crest that curls over the breaks in a plunging mass of water. These *plunging breakers* are a delight to surfers, who try to ride inside the curl and exit before being wiped out. The final type of breaker occurs on the steepest beach slopes. *Surging breakers* are long, low waves that cannot quite break before they meet the beach front. Instead, as they peak, their bottoms surge toward the shore, collapsing the about-to-curl crest. These three are the principal types of wave shoaling.

Wave refraction refers to the change in direction a wave makes as it nears the shore. Offshore waves are angled toward the coastline, but as they approach the shore, they turn so that they come in nearly parallel to the shoreline. This bending of waves in shallow water is influenced by the slope of the sea bottom and by the topography of the shoreline. Normally, the gradational contours of the sea-bottom slope

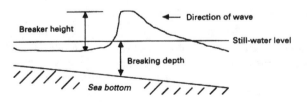

Figure 8-2   A schematic diagram of wave shoaling.

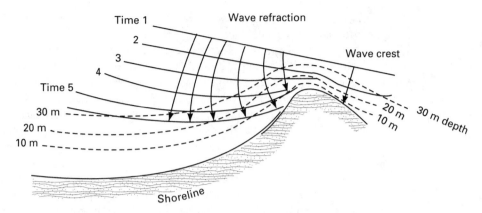

Figure 8-3   The refraction of a water wave approaching a headland. Portions of an advancing wave "feel the bottom" of the shallowing sea floor earlier than others.

parallel the contours of the coastline. In that case, the lower part of the wave feels the bottom first, then the rest of the wave catches up and the entire wave bends so that it strikes the coastline evenly. Where there is a headland or promontory, it is the advancing wave's crest that feels the bottom first, and in this case, wave refraction is focused on the topographical extension (Figure 8-3) and away from the broader adjacent regions. This concentration of breaking waves explains why headlands and promontories are so vulnerable to wave damage. Even in calm weather waves crash forcefully against these shores, and in stormy weather the breakers are enormous. Small

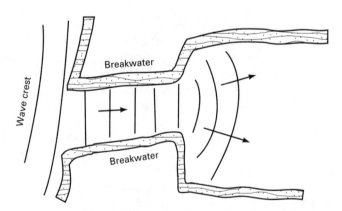

Figure 8-4   A map view showing the principle of water wave diffraction. Breakwaters compress the waves into an inlet. Then when the waves enter the wider inner harbor area, wave height diminishes and the waves expand laterally.

islands, too, are pregnable because refraction causes incoming waves to break all around their shores, no matter what the angle of direction of the offshore waves.

Wave diffraction is the modification a wave undergoes as a result of limitation of its lateral extent by a breakwater (a partial barrier erected offshore to protect a harbor from the full force of ocean waves). Only part of the wave strikes the barrier. The unobstructed part extends around (is diffracted by) the barrier and is compressed into the narrow inlet (Figure 8-4). Once this portion of the wave enters the wider area of the harbor or beach, its wave height diminishes and it expands laterally, focusing wave energy on the very area the barrier was meant to shelter.

This completes our introduction to waves. In the next two sections we will look at nature's most terrifying sea waves: tsunami and storm surges.

8-2 TSUNAMI

The most destructive waves on earth are not wind-driven but generated by a violent displacement of the ocean bed. **Tsunami** (the word is both singular and plural) are large long water waves caused by underwater earthquakes (seaquakes) or submarine volcanic eruptions or landslides. One of the strange things about these fearsome water waves is that they are hardly noticeable out at sea because their offshore wave height is insignificant. It is their *wavelength* that gives them their remarkable speed and energy. When one of these monstrous waves strikes a coastline, people standing onshore see the sea level rise above the highest tide level, then suddenly retreat far below the lowest water level—only to rear back and strike the shore at astonishing height and speed. The impact on a populated coastline can be calamitous. When a tsunami hit the coastal Sanriku district of Japan some 250 miles north of Tokyo in 1896, the first thing people noticed was the sea making a peculiar hissing sound as it was sucked out far beyond the limit of low tide. For a while there was a strange silence. Then, with a tremendous roar, the sea returned to attack, rising as high as 78 ft and surging inland to annihilate villages and drown 30,000 people. Less than an hour earlier an earthquake had shattered the seabed 700 km northeast of Sanriku—which shows the speed at which these waves can travel.

Tsunami (also called **seismic sea waves**) commonly occur in the Pacific Ocean because the majority of the world's earthquakes take place along that circum-Pacific belt known as the Ring of Fire (see Chapter 3). Most vulnerable are Japan (the word *tsunami* is Japanese for "harbor wave"), Hawaii, Alaska, Indonesia, and the Pacific coast of South America. A tsunami can strike any coastline, but because of their particular positions on the circum-Pacific belt (where 80% of all earthquakes take place) and the configuration of their coastlines (it is the angle of incidence relative to the shoreline that determines how bad the damage will be), these are the areas of the world most susceptible to devastating tsunami.

We do not know exactly how a tsunami forms, but Figure 8-5 diagrams a reasonable scenario. Tsunami are typically initiated by seaquakes (submarine earthquakes) that cause a sudden vertical offset in the sea floor. This abrupt offset of the seabed dis-

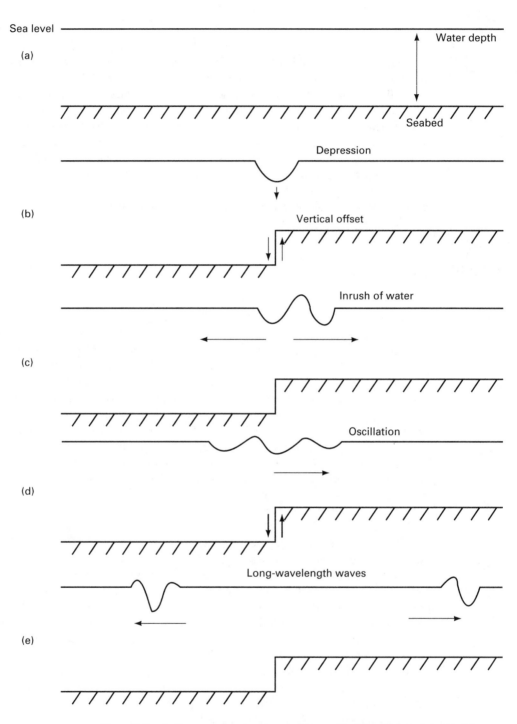

Figure 8-5 A diagram showing how a tsunami is thought to originate. (a) Sea surface before seismic disturbance creates sudden vertical offset (b), producing a depression, which (c) the surrounding water rushes in to fill. (d) As the water moves out at right angles, oscillation creates (e) long-wavelength waves that travel radially from the disturbed area.

places a large volume of water, producing a depression in the level of the sea surface. Water, being incompressible, rushes in to fill the depression and spreads out at right angles to the axis of the offset. Oscillatory wave motion is created, and waves of large wavelength move away from the point of disturbance.

The wave period of a tsunami varies from 20 minutes to several hours and its wavelength can be hundreds of kilometers. Both length and speed depend on water depth. For example, the average depth of the Pacific Ocean is 5 km (3.1 miles), so the typical speed of a tsunami at this water depth is 700 km/hr (435 mph). With an average wave period of 40 minutes, the length of a seismic sea wave (distance from wave crest to wave crest) will be 480 km (300 miles) and its height will only be several feet. This long wavelength and low wave height explain why ships on the open seas do not feel the passage of a tsunami. The Japanese fishing fleet out at sea during the 1896 Sanriku tsunami did not know anything unusual had happened until they returned to their wrecked port and found human bodies floating in the harbor waters.

As a tsunami approaches the shoreline, its speed, like that of all waves, decreases because of the slope of the sea bottom and its wavelength is compressed. The enormous amount of energy that had been stored in the very long wavelength is transferred to wave height, with terrible consequences. As wave height rises in shallow water—in some instances, to 100 ft or more—the tsunami curls over to strike with incredible fury. The most vulnerable coastlines for tsunami are along gulfs, bays, and estuaries. In these funnel-shaped inlets a tsunami can be amplified into a **water bore**—a vast wall of seawater weighing millions of tons that crashes inland with destructive power almost beyond belief.

One of the worst tsunami tragedies was occasioned by the great earthquake of Lisbon, Portugal, which occurred on November 1, 1755. The "calamity of the century" began in the morning with a noise witnesses described as sounding like grating thunder under the earth. This first shock was followed by an eerie silence. Then, with the second, much longer shock, columns and walls, roofs and spires, of shops, homes, and churches came tumbling down. After the third shock fires broke out, devouring a good part of the city. By this time the people of Lisbon were terror-stricken. But they had not yet faced the worst.

As aftershock upon aftershock reverberated through the city and a suffocating cloud of dust enveloped it, people rushed down to the harbor, intent on getting out by ship. There they were met by a towering tsunami about 50 ft high, apparently triggered by the sudden displacement of the sea floor off the coast of Portugal. The first wave was followed by others. In all, about 60,000 people were killed, many by drowning, others by fire. This tsunami raised waters along the European coastline and as far away as the West Indies; in Martinique and Barbados, where normal wave heights are 2 ft, the waves rose above 20 ft. Oscillations of landlocked bodies of water—**seiches** (pronounced *sāshes*)—were observed in northern Germany, Sweden, Scotland, and even Canada.

We said earlier that two other kinds of seismic disturbances can cause tsunami, though they do so less frequently than earthquakes. The first is undersea landslides and the second is submarine volcanic eruptions. The most notable volcano-instigated tsunami was that following the 1883 eruption of Krakatoa Volcano—itself probably the most violent natural explosion in 3000 years. Krakatoa is a volcanic island located in the Sunda Strait between Java and Sumatra islands of Indonesia (Figure 8-6). (Contrary to the wonderfully entertaining disaster movie about this explosion entitled *Krakatoa, East of Java*, Krakatoa is *west* of Java. By the time the Hollywood producers discovered this geographic fact, the movie title had already been cut—so East of Java it remained on celluloid.) The volcano was quiescent for two centuries until 1877, when a series of eruptions began. The worst started in May 1883 and continued through July and August. Krakatoa itself was uninhabited and people in the surrounding area had learned to take the spectacular explosions almost casually. Then on the morning of August 27 the whole northern and lower portions of the volcano were blown away in a last paroxysm. The resulting noise was heard at least as far away as Rodriguez Island, 3000 miles west-southwest in the Indian Ocean (Figure 8-7) and the enormous discharge of volcanic ash and dust darkened the skies for at least 150 miles around. Volcanic debris transported by air currents diffused for years over a latitude band from 30° N to 45° S, ultimately affecting North and South America, Europe, Asia, South Africa, and Australia. Spectacular sunrises and sunsets were observed around the world as atmospheric ash filtered solar radiation. In England Tennyson was moved by these brilliantly glowing skies to write:

> Had the fierce ashes of some fiery peak been hurled so high they ranged around the globe?
>
> . . .
>
> This wrathful sunset glared.
>
> *St. Telemachus*

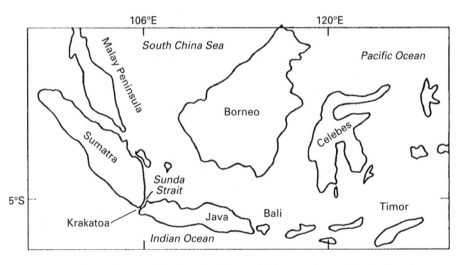

Figure 8-6   Location of Krakatoa Volcano. Krakatoa is located in the Sunda Strait.

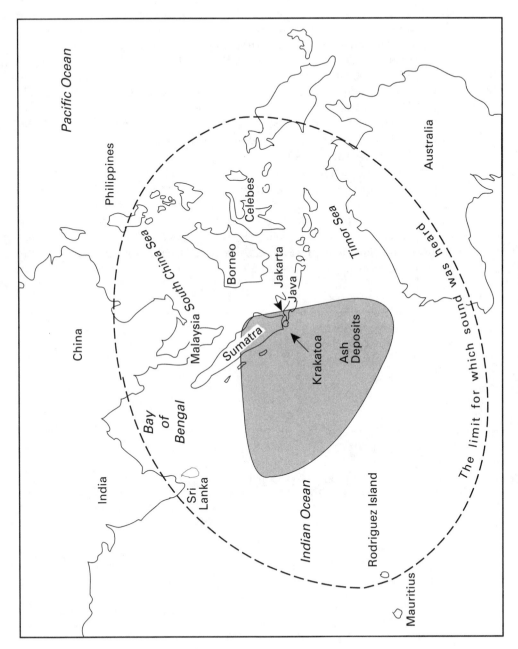

Figure 8-7 The noise range from the August 27, 1883, explosion of Krakatoa.

Now we come to the most serious consequence of the disintegration of Krakatoa: the generation of a long-period water wave (tsunami) and shorter-period but higher-amplitude sea waves (up to 50 ft). All along the coasts of Java and Sumatra colossal waves struck, inundating towns and killing 36,000 people. Racing acoss the seas, the tsunami ultimately reached Cape Horn (~ 8000 miles away) and the English Channel (~ 11,000 miles away). Associated with the Krakatoa eruption was a large atmospheric shock wave, and that could have induced tsunami waves at large distances from the explosion. Atmospheric coupling will occur if the speed of a shock wave is comparable to that of a tsunami wave in the deep ocean water. Oscillatory water wave motion was also observed in closed and semiclosed bodies of water at great distances from the volcanic eruption.

Figure 8-8 is a tide gauge record, pieced together from eyewitness reports of water-level changes, of the arrival and behavior of the tsunami at Jakarta, some 100 sea miles from Krakatoa. Beginning about 0200 local time on August 27, 1883, the water level slowly rose, culminating at 1215 in a vertical wall of water about $7\frac{1}{2}$ ft high. The water then rapidly fell 20 ft below the level observed at 0200 hours. The period of the wave from crest to crest was about 2 hours, and the disturbance lasted for at least 24 hours. Considering the arrival time of the tsunami, the distance it traveled, and the

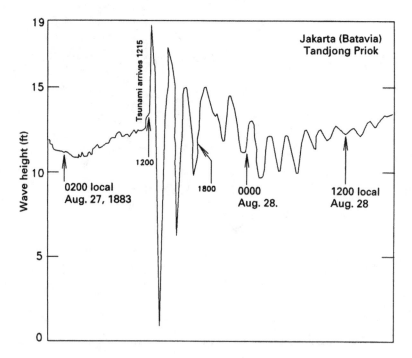

Figure 8-8 Tide gauge record showing the arrival and behavior at Jakarta of the tsunami generated by the August 1883 Krakatoa explosion. (Redrawn from Verbeek, 1886.)

average depth of the sea along its path, we can deduce that the onset of the tsunami occurred about $2\frac{1}{2}$ hours after the cataclysmic event at Krakatoa.

In 1883 there was no way to establish a system that could have detected the generation of a tsunami and given sufficient warning to save the lives of people in vulnerable coastal areas. But after a brutal tsunami swept Hawaii in 1946, an efficient monitoring system was set up to watch for signs of a tsunami every time a large earthquake is noted in the circum-Pacific belt. Strategically located observers in the United States, Japan, Taiwan, Chile, New Zealand, and other stations relay data to the network's center on Honolulu, which then alerts countries all around the Pacific. A tsunami travels at very impressive speeds, but distances are so great in the Pacific Ocean that this warning system works quite well. As Figure 8-9 shows, the transit time to Hawaii of a tsunami produced by an earthquake disturbance somewhere in the Pacific rim ranges from 5 to 15 hours. A tsunami warning system would not be as

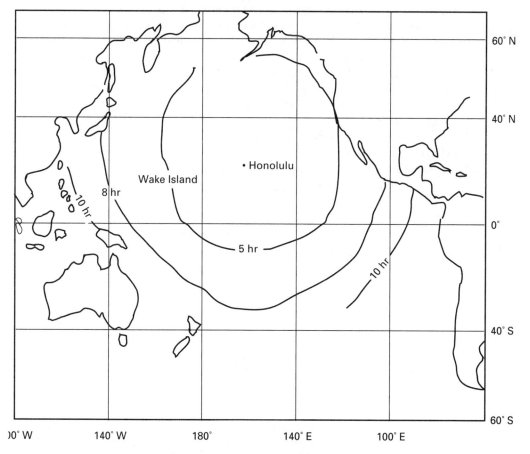

Figure 8-9   Approximate times it would take a tsunami to arrive at Honolulu from an initiating event on the Pacific rim.

effective in a smaller body of water, such as the Mediterranean Sea, because there the travel time distances would be much shorter.

Warning and evacuation are the only defenses we have against tsunami. The technology does not exist that can repel or contain these gigantic waves.

8-3 STORM SURGES

We touched on the subject of storm surges in Chapter 7 when we discussed the hazards of tropical cyclones. A *storm surge* is the transient motion, created by strong winds, of a portion of a body of water that results in a mean water level at the shoreline that is above the normal tidal level of oscillation.

On February 1, 1953, a strong storm surge struck the coast of the Netherlands, collapsing dikes, flooding 800,000 acres, and drowning 1800 people. Figure 8-10 shows the storm surge record made by a tide gauge at Rotterdam, from January 30 to February 2. The dashed line indicates the normal fluctuation of astronomically induced ocean tides for that period and the solid line indicates the observed level of water. Note that on February 1—the height of the storm surge when the protective dikes were breached—the observed level of water rose 2.75 m (9 ft) above high tide level. This storm surge was later described as a one-in-400-years event.

Several factors go into the making of a storm surge. First is *wind setup*, or the speed of the wind, its duration, and its direction. A strong, lasting wind that agitates coastal waters is an important precursor. Second is the *lowered atmospheric pressure* associated with approaching storms. This decrease in atmospheric pressure temporarily raises the local water level before the storm arrives. When the heralded storm finally descends, it adds to the already high water level and drives the rolling waters

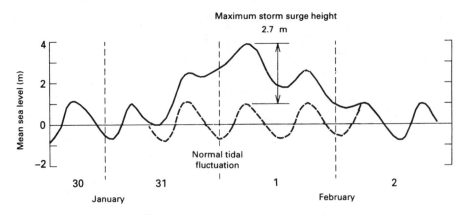

Figure 8-10   Storm surge tide gauge record at Rotterdam, the Netherlands, from January 30 to February 2, 1953. The height of the storm surge is gauged relative to the predicted astronomical tidal levels. Normal tides here reach a height of about 1.2 to 1.3 m.

toward the shoreline. Once the storm moves on, the *slope of the sea bottom* controls the height of the surge: Shallow seabeds produce the highest waves. *Topology*, or configuration of the shoreline, also plays an important role. Low-lying coastal areas that focus wave energy are most at risk for disastrous storm surges. The final factor is timing: Storm surges that strike at *high tide* are always the most devastating. As we saw in Chapter 7, all these factors operated to create the tragic cyclonic storm surge of 1970 in Bangladesh.

Since most of the flooding, and most of the fatalities, caused by tropical cyclones and other storms come not from the intensive rains of the storm itself but from the storm surge that succeeds them, it would be very useful if we could predict when and where surges will happen. Two of the observable facets of storm surges are their height and frequency of occurrence. In some parts of the world data on these features are regularly collected and have been for some time. Can we use these to infer coming events—is the past key to the future?

Figure 8-11 is a graph of the height of storm surges in the North Sea from 1883 to 1953 plotted against their recurrence intervals. Note that the plot is semilogarithmic: The arithmetic scale of surge height in meters is plotted against the logarithmic scale of recurrence time in years. One of the important characteristics of the temporal variations in *magnitude* of natural hazards such as storm surges, and earthquakes, is that they tend to have a linear relation when plotted on a semilogarithmic plot. The data in Figure 8-11 suggest that a 3.3-m-high storm surge will occur every 70 years, a 4-m surge every 800 years, and a 5-m surge about every 10,000 years. The Dutch have, accordingly, raised their dikes to withstand the most catastrophic (5-m) storm surge predicted.

We need to look at such data plots with a critical eye, however. Remember, the data for the North Sea cover a period of only 70 years. Can we extrapolate that a 5-m surge will take place but once in 10,000 years from a 7-decade sample of data for smaller events? Generally speaking, can smaller-size events be used to predict the occurrence of larger events? Since the time interval for which we have reliable records is always small, any undocumented event in the historical past could badly skew the slope of the graph used to infer future events.

Finally, this sort of plot is helpful for estimating the *average recurrence time* and size of a "big" event—but *not its timing*! For if a large, rare event occurs today, there is nothing in a statistical sense that would preclude another disastrous event of the same type from happening tomorrow. The long-term average may not change, but there are sporadic or cyclical clusters of events over smaller intervals of time—which is both intriguing and perplexing. It is human nature to assume after one rare disastrous storm surge that another of such magnitude is not "due" for a long time. This is a dangerous assumption.

Short-term forecasts of events (tropical cyclones, other massive storms) that have a strong probability of precipitating a storm surge are quite good at this point—if they are heeded. (In 1970 the government of Pakistan—which then controlled Bangladesh—did not issue an evacuation order to vulnerable populations until the evening the cyclone struck, although it had received strong warnings about the

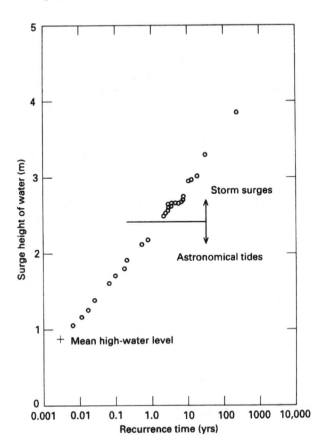

Figure 8-11   Surge height of water in the North Sea from 1883 to 1953 plotted against the logarithm of the recurrence time in years. (Data annotated and taken from Wemelsfelder, 1961.)

approaching cyclone. Hundreds of thousands of people perished without a chance at survival.) The system is not flawless, so false alarms go out, which sometimes makes people cynical, but there is no doubt that monitoring of storms and timely warnings of possible storm surges have saved many lives in recent years.

What about containing storm surges? Although no human device can hold back a large tsunami, storm surges of all probable magnitudes can be defended against where there is the will and the wherewithal. The Dutch have their dikes (raised considerably since the 1953 disaster) and the Japanese their seawalls; both are expected to stand against the foreseeable worst. After the 1900 catastrophic storm surge in Galveston (Chapter 7), a 17-ft stormwall was built, which furnishes some protection, though many believe it can be breached by the storm surge of a strong hurricane. Other coastal areas in the world are far less protected, either out of apathy or poverty. Bangladesh, for instance, remains as much in jeopardy as ever.

8-4 ICE HAZARDS IN THE OCEANS

On April 15, 1912, the *Titanic*, the largest ship afloat at the time, left on its maiden voyage across the North Atlantic. The drama is well known: The liner collided with an iceberg and 1500 people out of a total of 2200 passengers and crew lost their lives. After this tragedy an International Ice Patrol was established to track the movement of icebergs and warn ships of their location.

What is an **iceberg**? Huge glaciers (very slowly moving masses of freshwater ice) cover Greenland toward the North Pole and Antarctica at the South Pole. At the edges of these glaciers, over bays and other sheltered waters, a permanent ice shelf forms that seasonally breaks off into the ocean as chunks of ice, or icebergs, which float on the open seas. These menaces loom 60 m (196 ft) above the sea surface (and several times that below the surface) at their point of origin, though they gradually wear away to maybe one-tenth that size by the time they reach warmer latitudes. Icebergs rarely move farther south than 40° N in the northern hemisphere or farther north than 40° S in the southern hemisphere, though in 1894 an Antarctic iceberg was reported up near the Tropic of Capricorn ($23\frac{1}{2}$ ° S)! The year the *Titanic* sank there was an extraordinary number of icebergs in the North Atlantic. Even in ordinary years icebergs on the open sea are significant hazards to shipping, fishing, and offshore oil drilling.

Icebergs are formed from freshwater land ice. Sea ice also produces oceanographic hazards. There is a permanent cover of sea ice 3 to 4 m thick over 70% of the Arctic Ocean in the north and encircling the continent of Antarctica in the south. For much of the year solid and floating sea ice is also found around the numerous northern islands of Canada and Russia, the islands of Greenland and Iceland, and the oil-drilling areas off Norway and Scotland.

Icebreakers—ships equipped to cut a channel through icebound waters for other ships to follow—make passage through such waters as the North Sea, Barents Sea, Norwegian Sea, and Arctic Ocean possible during part of the year. When ocean currents and winds form *hummocks* (pressure ridges) on the sea ice, however, the ice wins: Even icebreakers are of no use.

Pack ice is pieces of floating sea ice driven together into a single treacherous mass. It occurs in the more closed bodies of water in the higher latitudes, such as the Bering and Baltic seas, the seas of Japan and Okhotsk, and the Gulf of St. Lawrence. When pack ice drifts toward shore, it can block harbors during their usual ice-free season. Like icebergs, drifting pack ice threatens offshore oil-drilling platforms.

REVIEW

1. Why does a change in the speed of a water wave approaching a shoreline produce an increase in hazard?
2. What causes waves in the open ocean?

3. Why is the shape of a coastline an important factor for assessing potential damage from waves?

4. What is fetch?

5. What is the typical wave speed of a tsunami?

6. Why is a tsunami not felt by ships at sea?

7. A tsunami was generated by the violent explosion of Krakatoa Volcano. Why?

8. What is a seiche?

9. Explain how the tsunami warning system established in Honolulu operates.

10. Suppose that between 1900 and 1980 two storm surges with heights of 1.2 m were recorded at the Florida coastline. Next, suppose that five storm surges, each with a height of 0.6 m, were observed there within this same period. Calculate how many times a 1-m surge should have occurred within this time frame.

11. What are the chief factors in the formation of storm surges?

12. Why should attempts to forecast future storm surge heights based on recurrence data from the past be looked at critically?

9

River Floods

On a holiday we were driving along a river embankment of the Mississippi when we heard over the radio that a big storm was approaching. We were not overly alarmed because we knew there was a system of movable **weirs**, or dams, designed to back up the flow if the river rose too high from the rains. We also knew there were tide gauges positioned to register the water's rise and forewarn of any flooding with remarkable accuracy. So we proceeded down the levee confident we would receive a timely radio warning if the river was expected to overflow.

Were we surprised! The weirkeepers must have been asleep at the switch because the weirs had not been opened and a surge of waters broke through the barrier. As we drove along, wondering why no warnings had been given on the radio, we hoped the floodwaters would not top the levee. We were lucky that day—the levee held and we got away safely. The engineers' scheme of river control worked, despite carelessness on the part of some of its operators. But no water-control system yet devised can withstand the strength and power of a mighty river in full flood, as Americans found out during the Great Flood of 1993.

Floods are the most widespread natural hazard in the world. There are two types: inundation by the sea, which we covered in the last chapter, and overflowing of inland waters, especially rivers, which is the subject of this chapter.

Human beings need water badly, so from ancient times they have often chosen to live by great rivers. The same river that supplies drinking and washing water, fish, and a valuable transportation route provides, in its floodplains, fine grazing and cropland. Sometimes, however, the river's bounty turns into disaster. The Great Deluge—an enormous flood that destroys the whole world and nearly everyone in it—is an episode in most of the world's mythologies. So many legends, widely separated in time and place, prove that the fact (and fear) of floods has been great all over the earth.

9-1 THE BIBLICAL FLOOD: MYTH OR HISTORY?

The most famous version of a Great Deluge is that found in Genesis (Chapters 6–9). For 40 days and nights, the Bible says, "the waters prevailed, and were increased greatly upon the earth . . . and all the high hills, that were under the whole heaven, were covered. . . . All in whose nostrils was the breath of life . . . died" (Gen. 7:19–22). All, that is, but Noah and his household, who, warned by God, stayed safe and serene inside the ark that God had commanded to be built. Also within the ark were a male-female pair of every species of beast and fowl, taken aboard so that the earth could be repopulated after the flood ceased. The biblical language suggests a catastrophic submergence of the entire world—an event for which we have no geological, botanical, or zoological evidence—but it must be remembered that to the ancients the "whole world" was *their* world. A flood of great magnitude in Mesopotamia, that great fertile plain drained by the Tigris and Euphrates that formed the heartland of several ancient civilizations (Sumerian, Babylonian, Assyrian, Chaldean), would seem like a world-wide deluge.

There is some archeological evidence that more than 4 millennia ago this area may have been overwhelmed by a gigantic flood, one greater than anything the region has seen ever since. Moreover, there is a version of the Great Deluge that is much earlier than Genesis. It is contained in the Babylonian Epic of Gilgamesh (~ 2000 B.C.) and it bears a strong resemblance to Noah's story. Here, too, the flood is divine judgment on human sin, and one good man, Utnapishtim, is singled out to be saved. Like Noah, Utnapishtim is commanded (by *gods*, in this case) to construct a great ship to shelter his family, servants, and animals, though in the Babylonian version the flood lasts only a week.

If, as seems likely, there was an unparalleled flood in Mesopotamia that destroyed all the habitable land and drowned most of the people living on it, what natural causes could lie behind these religious accounts? Contemporary speculation has produced several theories. One proposal is that persistent torrential rains caused the Tigris and Euphrates to overflow their banks. The problem with this theory is that the average rainfall in Mesopotamia always seems to have been low. Another suggestion is that a local reservoir broke, but it is not plausible that this could cause the kind of flood described in Genesis and Gilgamesh, even allowing for much exaggeration. Perhaps the most reasonable natural-origins hypothesis is that either an earthquake or a typhoon loosed a tsunami that swept into Mesopotamia from the Persian Gulf. Notice, though, there are common themes in both ancient descriptions: condemnation of a "sinful" world, divine *forewarning*, the sheltering of a select few, and then the Great Deluge. Is it possible that the disaster was triggered by human beings for religious-political-military reasons?

There is no doubt that human intervention (for whatever reasons) has often exacerbated climatic events. Southern California started to burn in the fall of 1993 because hot desert winds struck the tinderbox of drought-dried vegetation, but within a short time arsonists were setting fires that rivaled anything the Santa Anas had wrought. In recorded history it is not unusual for a conquering "barbarian" army of nomads to destroy works of a "civilized" settled community that is seen as corrupt and irreligious. Ancient Mesopotamia had a prodigious civilization based on sophisticated agriculture, with irrigation works, dams, canals, and embankments to mitigate the effects of any heavy flooding. Perhaps enemies intent on destroying this civilization started a great flood (or enhanced one in progress) by wrecking these elaborate works (with some inside help?). Such an event, shaped and embellished by writers with a theological message, could lie behind the biblical account of Noah's flood.

9-2 WHY DOES A RIVER FLOOD?

All rivers flow finally into the sea. The land area that is drained by a river and its tributaries on the way to the sea forms its **drainage basin**. A river's size is proportional to the size of its drainage basin and the amount of rain deposited over that basin. The Amazon in South America, the river with the greatest volume discharge in the world, has a huge drainage basin (over one-third of the continent) and flows through dense tropical rain forests.

Rivers vary greatly in their discharge rates, particularly in regions subjected to heavy seasonal rains. In these regions a river will typically flood in the rainy season, and then dry to a trickle during the rest of the year. In more temperate climates, rainfall and, consequently, river discharges are more evenly distributed.

Like coastal flooding, river flooding is primarily caused by atmospheric effects (Figure 9-1). Snowmelt and ice melt can produce river flooding, but the more common cause is colliding weather fronts that become stuck over the drainage basin. Day after day there is a slow, inexorable rise in the water level as the relentless rains tax the river and its tributaries beyond their capacities. The type of ground on which the rain falls plays an important part. If the ground is firm, rocky, relatively impermeable, and barren of vegetation, water will rush downward to the tributaries and the main drainage river, rapidly raising them to flood levels. High floods are also generated by heavy rainfall on ground that is already supersaturated by previous rains, especially in winter when surface evaporation is low. On the other hand, flooding will be somewhat slower if the soil is permeable and has vegetative cover, for both the ground and the trees will soak up some of the moisture.

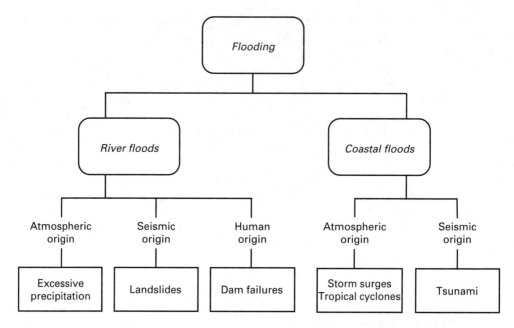

Figure 9-1 The causes of river and coastal floods. Atmospheric effects, which produce large amounts of rainfall, are the most important contributor to flooding.

9-3 FLOOD FREQUENCY

Flood frequency data have been collected for only a short historical time, ranging from decades to at most a century or so, but, as with earthquakes and storm surges, these data are useful for long-term predictions of unusually great events. The procedure for deriving a flood magnitude-frequency of occurrence relation that enables us to estimate the recurrence times of great floods uses a probability model, so readers unfamiliar with that branch of mathematics might want to turn to Appendix A first.

The procedure is as follows. Flood discharge data for a drainage area for N years of record are ordered from the highest ($M = 1$) to lowest ($M = N$) rank on an annual basis. The series forms ($N + 1$) rank classes. The probability of a random event of magnitude or size (in this case, discharge in cubic meters/second) being equal to or greater than an event of rank M is

$$P(x) = \frac{M}{(N+1)}$$

and the mean return time of each flood is simply

$$T = \frac{(N+1)}{M}$$

This probability model is based solely on rank orders and is independent of a probability distribution. If, for example, the length of the flood discharge record is 20 years, the largest discharge is expected to have a return time of 21 years, the next highest discharge ($M = 2$) a return time of 10.5 years, and so on. These data are usually plotted on graph paper that compresses the upper end of the time scale and expands the lower end, thus emphasizing the more frequent recurrence intervals of smaller-discharge events (Figure 9-2).

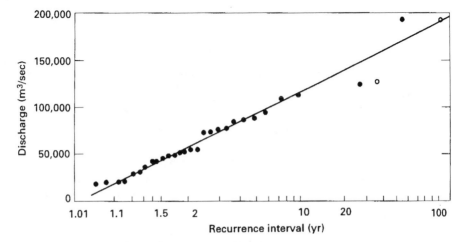

Figure 9-2   Peak flood discharge curve versus recurrence interval in years for the Eel River in California. (Data are from U.S. Geological Survey.) The solid circles cover the period from 1932 to 1959 and the open circles are for two historical events outside that time frame (1915 and 1862.)

Flood data allow a *statistical* estimation that a flood of a certain size (cubic meters of discharge per second) will recur within a certain period of time. The reciprocal of the recurrence (return time) gives the probability of occurrence of a flood of that size. For example, a flood peak with a 100-year return time has a probability of 0.01 or 1% of being exceeded in any given year (Figure 9-3). Once again, we must caution that all such analyses are questionable for predictions into a future time period exceeding the length of the historical record. So when it is said—as it often was in the popular media—that the Great Flood of 1993 in the U.S. Midwest was a "once-in-500-year flood," it should be remembered that (1) statistically, there is nothing to preclude another flood of that magnitude the very next year and (2) our reliable record of floods in this floodplain is much briefer than 500 years.

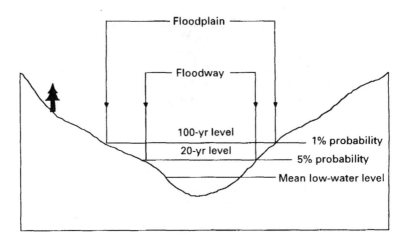

Figure 9-3   A diagram illustrating the concept of a 20-year or 100-year flood with its associated probability of being exceeded in any year.

9-4 CHINA: LAND OF CATASTROPHIC FLOODS

Catastrophic floods are sudden, of great magnitude, and extremely destructive to life and property. Of all the countries in the world, China has historically suffered the greatest loss of life from flooding. Two principal rivers are responsible for these devastating floods.

The more northerly of the two, the Hwang-Ho, is commonly known in English as the Yellow River for its deposits of wind- and water-borne *loess*, a yellowish-brown loamy material. To the Chinese the river is known as "China's Sorrow," "The Ungovernable," and "Scourge of the Sons of Han" for its lethal floods. The Hwang-Ho has the distinction of being the most flood-prone river in the world. For 4 millennia the Chinese have been attempting to control it with dredging and engineering works, but periodically the river surges over its embankments and engulfs the countryside. In the year 1887 the river rose so high that it topped 21-m (70-ft) levees and inundated 28,000 km² (50,000 square miles) to depths of 9 m (30 ft). More than 900,000 people died in this tragedy and another 2 million were made homeless as the waters devoured hundreds of villages. This was one of the very worst river floods in history.

Another of the great Hwang-Ho floods was humanmade. In 1938 Chiang Kai-shek tried to stop the Japanese invading army by dynamiting a hole in the river's southern levee. The ensuing flood did stymie the enemy, but it was an incredible disaster for the Chinese: Thousands of villages were wiped out, perhaps half a million peasants drowned, and 6 million lost their homes. The full death toll from this deliberate inundation will never be known, but it is estimated that several million Chinese died in the famine caused by the flood's destruction of agriculture.

China's Sorrow is also China's Bounty, for this great river has an immense floodplain with rich farmland that has fed much of the country for thousands of years. (A **floodplain** is that extent of relatively flat land adjacent to a river that is subject to flooding when the river overflows. The deposit of river sediment left behind after the water recedes is a fine fertile soil.) An extensive modern system of dike construction was begun in the 1950s, and there have been no devastating floods in recent decades. Yet the long-term prognosis for controlling the Hwang-Ho is not good. The river is constantly raising its level by deposition of loess at a rate of at least 0.3 m (1 ft) every 100 years. The river is now ~ 20 m (~ 60 ft) higher than its neighboring floodplain, and a flood of large magnitude could breach the protective dikes and levees. Another problem is that the Hwang-Ho **meanders**, and this tendency to make wide swings in its route facilitates the overflowing of dikes. The Chinese are attempting to deepen the river's channel and straighten its course to the sea, but this is an immense project for a 3000-mile-long waterway.

Unlike the Hwang-Ho, China's other problem river is an important artery of commerce. The Yangtze Kiang (or Chang Jiang, as it is now called) flows southeasterly to the sea through a humid region subjected to high seasonal rainfall. Devastating floods were recorded for this river in 1153, 1227, 1520, 1560, 1788, 1796, 1860, 1870, and 1931. The 1931 flood was an appalling episode that affected 60 million people, with a death toll in the millions. By 1954, when the next great flood of the Yangtze took place, the Chinese were better prepared. Though flood levels reportedly reached 30 m (~ 100 ft), by raising the dikes and other feats of human perseverance, they managed to avert a disaster.

9-5 THE GREAT MIDWESTERN FLOOD OF 1993

In the summer of 1993 the Mississippi and its tributaries overwhelmed levees, erased towns, knocked out bridges and airports, and rearranged the landscape of good portions of the Middle West for some time to come. Americans watched, stunned, as week after week the television cameras captured the spectacle of inexorably rising waters, people's desperate efforts to barricade their homes and towns by heaping sandbags, concrete, dirt, and rock upon the levees, and the numbed defeat on their faces as, again and again, the river overmastered their work. By the end of the summer, the flood had spread across 354 counties in 9 states, inundating 13.5 million acres, displacing at least 50,000 people, and causing 50 fatalities.

The Great Flood of 1993 started in April in Iowa, when the ordinary spring rains refused to let up. All spring and into August parts of the Midwest received as much as 10 times the normal amount of rainfall, which swelled the Mississippi and its 50 tributaries (including the mighty Missouri and Ohio rivers). The Mississippi is the largest river in the United States, draining 1.25 million square miles. Ordinarily, when its waters rise, there is a backflow into its tributaries. These waters then rise higher than normal and overflow their banks and levees in local floods of small or medium magnitude. What made the 1993 flood such a disaster was that the unceasing rains caused most of the tributaries to be in full spate simultaneously—the first time this had happened in recorded history. There was nowhere for the waters to back up, so they flowed down in a slow, inexorable tide, overtopped the embankments, and reclaimed the floodplain. Where farms and towns had flourished, there were huge lakes. Rescue boats cruised among the rooftops, picking up the hapless stranded.

The Mississippi had flooded badly before, and over the last century a grand scheme of 28 locks and dams and 6000 miles of levees had been constructed to keep the river and it tributaries manageable and navigable. (The Mississippi moves 2 billion tons of goods a year and transportation on its waters links markets from 40% of the United States.) The locks and dams held up well, as did almost all of the levees built by the federal government. But most of the levee system on the Mississippi floodplain was constructed by private interests and municipalities, and 70% of these levees failed. They could not control the wild river channels cut by a flood of this magnitude.

What went wrong with the grand scheme? Some complained that the locks and dams on the Upper Mississippi were operated improperly, but it is doubtful that a different judgment call on when to hold and release their waters would have made much difference to the outcome. Others pointed out that the magnificent engineering system had actually made things *worse* by channeling the cresting waters too rigidly, so that they gathered ever more force along their route. It may be that we just have to face the fact that flooding of floodplains is a natural process and dismantle some of the levees to allow the "green-belting" of parts of these areas. Much of the land would then be reserved for agriculture rather than for residences, commerce, and industry; other, more marginal land would become wetlands. Considering that 7 million people live along the Mississippi, this is easier said than done.

9-6 FLASH FLOODS

A thunderstorm in the wrong terrain can produce a sudden violent flood that vandalizes a sizable area. When heavy rain falls over a locale that drains into steep trenches, these quickly choke, forming torrential streams that rage for miles with astonishing speed. Its climate and topography make the American Southwest particularly vulnerable to these **flash floods**. Central Texas, dry most of the year, is seasonally subjected to harsh storms coming in from the Gulf of Mexico. Then its steep drainage slopes make it prone to flash floods. Utah, Colorado, Arizona, and Southern California are other areas whose terrains make them susceptible to flash floods during sudden brief storms.

It is not just the water that gives flash floods their destructive power. As the torrent pours through gullies and small valleys, it picks up masses of earth, rock, and other debris. The flood then turns into a heavy mudflow that slows and spreads out, pushing boulders, trees, cars, and even houses before it. Flash floods can be particularly hazardous in urban areas where much of the ground is paved over and the drainage channels are fixed (Figures 9-4 and 9-5).

The worst thing that can happen in a flash flood is for a dam to fail. In that case,

Figure 9-4 Flash flooding in Santa Ana, California, on February 25, 1969. The causative storm forced the evacuation of 6000 people in seven southern California counties. (Courtesy of Orange County Register.)

Figure 9-5   Urban flooding damage at Glendora, California, on January 22, 1969, caused millions of dollars' worth of damage and took 29 people's lives. Observe the amount of sedimentary debris transported.

the flood volume, in a matter of hours, can exceed the discharge rate of some of the world's mightiest rivers. In June 1972, in the Black Hills of South Dakota, over 1 m of rain fell in 6 hours. A dam broke and a wall of water engulfed Rapid City, leaving 230 people dead and 2900 injured. Property damage, including 1500 destroyed cars, was estimated at $90 million. A flash flood of this magnitude has a statistical probability of occurring once in 2000 years.

Dams are not supposed to fail, but often they are not properly maintained, were poorly designed to begin with, or were inadvertently positioned on an unstable geological fault. *Natural* dam failures like the sudden bursting of ice-dammed lakes or glacier bursts (called *jökulhlaups*), can turn a flash flood hazard into a great disaster. In November 1985, 20,000 people were killed by a glacier burst from the Nevado Del Ruiz volcano in Colombia, South America.

The most important factor in flash floods is not the size of the water channel but its peak discharge of floodwaters compared to its mean annual discharge. Damaging floods are those that produce discharges many times higher than the mean. Small streams have a high flash-flood potential if they are in an area of steep slopes, thin soil cover, and intense local rainfall.

9-7 BENEFICIAL FLOODING: THE NILE

For thousands of years Egyptian harvests depended on the annual flooding of the Nile, the second-longest river in the world (after the Amazon). The peasants would trudge into the rich slime left behind by the flood to plant their crops, then irrigate them through the dry months until the harvest season.

Knowing basic astronomy, the priests of ancient Egypt were able to calculate the times for the Nile's annual rise and fall. These were regular, but sometimes the river rose too high, generating disastrous floods. So thousands of years before the beginning of the Christian era pharaohs constructed dikes along the river's banks and directed when they should be opened for beneficial flooding of the agricultural floodplain. Records on the height of the Nile River were etched in stone at least as early as 1750 B.C. (some historians claim documentation goes back to about 3600 B.C.). The dikes were opened to flood the lowlands when the river reached the level of 16 cubits (1 cubit = 52 cm or 0.5 m). At 18 cubits the middle elevations were flooded, and at 20 cubits the high ground. Every year when the Nile reached 16 cubits the Egyptians held a festival called the *Wafa* to celebrate the renewal of the parched land.

Until recent times the fertility of Egyptian and north Sudanese agriculture depended solely upon the flooding and irrigation water from the Nile, and even today both countries rely heavily upon the river. As with the Hwang-Ho, however, siltation is a growing problem. The entire Nile valley has been rising ~ 13 cm per century because of sediment deposition. So any significant increase in the amounts of water supplied to the Nile could result in bad floods, like the ones in 1874 and 1878, when the waters rose 9 m at Cairo.

9-8 PREVENTION AND MITIGATION

River flooding has always been a global hazard because great numbers of people have always congregated on floodplains for their fertile soil and for the commercial advantages of a navigable river. What has changed in modern times is the degree to which the valleys of rivers and streams that drain away normal floodwaters have become crowded for industrial reasons and because of population pressures. We have altered the floodplains to suit our needs to an unprecedented extent. By paving over increasing amounts of the earth's surface, we are minimizing natural drainage and enhancing the opportunities for great floods. Where once a flood of great magnitude could rage with little consequence except, perhaps, to agriculture, today it can wreck the infrastructure of civilization.

At least the hazard is now being seriously recognized, and that is an important first step in mitigating the problem. Others are accurate risk assessment, reliable fore-casting, warning and emergency response systems, and sound flood-control pro-grams. In addition, we also need to consider making a greater effort in land use con-trols.

Risk assessment for river floods involves an evaluation of the likelihood that a region will be inundated to a certain depth. The depth of a 1-in-20-year flood is esti-mated and plans are made accordingly. This does not mean that a flood of the speci-fied depth will definitely occur once every 20 years but that, *on average*, one such flood is to be expected every 2 decades. Prudent planning allows for this average recurrence interval.

Flood forecasting and warning systems are essential for minimizing loss of life and property damage. Centers have been set up in the principal river basins of the United States to monitor dangerous river conditions and to warn people of coming floods. The chances of saving lives rise in proportion to the length of the warning time, and these systems have been quite reliable in their predictions and warnings.

Emergency response performance has been more checkered. The response of the Federal Emergency Management Agency (FEMA) to the Great Flood of 1993 was creditable: In a flood of this magnitude, it is important to have a nerve center that can coordinate actions by federal, state, and local governments to avoid duplication in one area and inaction in another, and by and large, FEMA managed this very well. Local and state emergency response systems in flood-prone areas have ranged from excel-lent to inept in recent years. It is crucial that emergency personnel have a clear under-standing of what kind of flood might happen, where it is likely to take place, and how quickly it will develop. They must also have well-trained workers, adequate supplies, and workable delivery systems.

Flood control has been practiced by advanced civilizations for several millen-nia, though total control is still impossible. One effective conservation measure is to plant trees and deep-rooted grasses on natural drainage slopes to absorb water. Another solution that has been implemented in many parts of the world, often in con-junction with the first, is flood-control reservoirs. If there is a closely spaced second flood, however, it may not be possible to draw down the filled reservoirs in time to accept the new onslaught of floodwaters.

Sedimentation, or siltation, is a problem in many rivers. As constant deposition of sediments builds up the river bottom even a minor amount of excess water will make the river overflow. Dredging to deepen the river channel is an expensive but not impossible solution.

Our most elaborate systems of flood control use dams, locks, dikes, diversion channels, and levees to bend rivers to our will. These protective systems have allowed us to expand agricultural land, improve transportation, and build communi-ties—sometimes where we should not. The Great Flood of 1993, which had such a transfiguring effect on the land, is also transforming our thinking about flood control. For one thing, there are economic limits to what we can do. The relentless develop-ment of the floodplains has been canceling out our flood-control programs; it is impossible to keep up with population spread into ever-more-marginal lands. For another thing, the 1993 flood demonstrated how the most ingenious schemes for impeding the natural flow of a river in full flood can exacerbate the situation. Every containing dam, every levee to prevent overflow of the riverbanks—every obstruction we put in the river's way—raises the level of the waters behind the obstruction, and therefore the potential flood level. Removing some of these human-made obstructions would increase the discharge capacity of river channels.

Given the great demand for housing and industrial development, *land-use controls* are controversial the world over. But it is time to acknowledge that we cannot do without them. Housing and industrial construction should not be permitted in areas that have been zoned as susceptible to, say, a 1-in-10-year severe flood. Elsewhere it may be cheaper to buy up farmland and use it as a natural floodplain than to build (or rebuild) levees. These areas could become nature preserves with perhaps limited hunting and fishing allowed. We also ought to have and enforce environmental laws against the dumping of mining refuse in rivers, mill sluice gates, excessive fish traps and weirs, and overwide bridge piers—all of which obstruct water flow and raise the likelihood of flooding.

Flood prevention and mitigation require that we both spend for flood control and slow or stop the development of the floodplains. The bottom line is that it is more important to protect communities rather than open farmlands developed on *natural* river floodplains. Reconstruction must focus on future protection to minimize future risk yet maximize the long-term viability and economic development of flood prone regions.

REVIEW

1. What do the accounts of an ancient Great Deluge found in the Bible and the Epic of Gilgamesh suggest in their similarities?
2. What are the two primary causes of river flooding?
3. If a flood peak height has a 20-year return time, what is the probability that this height will be exceeded in one year?
4. If a particular flood peak height has a 50% probability of being exceeded in one year, what is the average or mean return time of this flood height?
5. Why is the Hwang-Ho River constantly raising its level?
6. Describe the genesis and development of the Great Flood of 1993.
7. Why is flash flooding often so severe in urban areas?
8. Name two beneficial aspects of river flooding.
9. List the five major steps in an effective program for the prevention and mitigation of river flooding.

10

Some Accident Scenarios

Now that we have finished describing the major natural hazards, it is time to consider how to determine whether we should attempt to lessen a particular hazard's risk, learn to live with it, or avoid it altogether by choosing to live elsewhere. Before we set up some scenarios that illustrate how to determine risk, let us look at a real-life hazard whose risks were underappreciated until it became a disaster.

10-1 BHOPAL

In 1984 a major industrial accident occurred in Bhopal, India, at a plant built to manufacture pesticides designed as a quick fix for Third World countries plagued by the infestation of crop-damaging pests. An agricultural "green revolution" was envisaged in which the pesticides would improve crop productivity and alleviate chronic food shortages. Unfortunately, the plant developed a poison gas leak and somewhere between 2000 and 10,000 people in the surrounding area died of pesticide poisoning and many more became gravely ill. This, of course, was a human-made disaster, but the lessons we learned from it are directly transferable to natural hazards. The first is that understanding a hazard and anticipating its effects are crucial to hazard assessment.

The postmortem studies of the Bhopal tragedy focused on four aspects of risk assessment of hazards: (1) cost-benefit analyses; (2) the establishment of a safety threshold; (3) the development of a causal hazard chain; and (4) the consequences of not making decisions. All these aspects are applicable to the assessment of risks from natural hazards. We discussed the setting of safety thresholds in Chapter 1, and cost-benefit analyses and the consequences of not making decisions are self-explanatory, so let us take up the development of a causal hazard chain.

10-2 CAUSAL HAZARD CHAINS

A **causal hazard chain** is a kind of accident scenario—a predetermined sequence of hazards and potential failures, or disasters, each of which has a mitigative alternative. Theoretically, a causal hazard sequence could be infinite, but as a useful assessment tool, it must be limited to plausible failures based on the best data. When we limit a causal hazard chain, then, we are saying that this chain is susceptible to this predetermined set of failures and no others. The danger here is obvious: We prepare mitigations for *only* those failures we have anticipated.

Causal hazard chains are necessary for developing a decision tree in which finite probabilities are assigned to various risks (links in the chain). Figure 10-1 is an example of a causal hazard chain. At each link in the sequence there is an alternative that might prevent the end result—death. Notice, though, that these alternatives are not unequivocal because they might not be available at the correct time in the sequence.

An often overlooked element of risk evaluation is the different attitudes individuals have toward risk. Some people are so risk averse that they want to avoid possible extreme consequences at all costs. Others are prone to take risks, and still others are

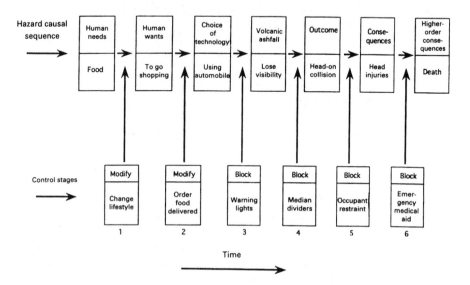

Figure 10-1   A causal hazard chain. (Modified from Bick, Hohenemser, and Kates, 1979.)

indifferent. To test your own attitude toward risk, consider the following proposal. Suppose that there is a lottery with only two outcomes, each of which has an equal likelihood of occurring: Either you win $100,000 or you win $500,000. Your chances of winning either sum of money are exactly the same. Now suppose that you are offered $300,000 if you decline to play the lottery. Would you choose the expected outcome of a sure $300,000 or would you choose to gamble on the 50/50 chance that you could walk way with $500,000?

If you selected the expected outcome rather than the risk choice of the lottery, you are risk averse. Conversely, if you chose to play the lottery, you are risk prone. Indifference to the question—having no preference between the lottery and the expected outcome—marks you as indifferent or risk neutral. But what if the numbers in this scenario denoted human lives rather than money, so that the choice is between playing the lottery with an equal chance of ending up with 100,000 and 500,000 fatalities or declining to gamble and ending up with a sure outcome of 300,000 fatalities? Would you answer the question the same way?

The concepts of risk aversion, risk proneness, and risk neutrality are vital to assessments of natural hazards. Risk aversion is synonymous with the strong desire to avoid a catastrophe. It means preferring a 10% chance of 10,000 fatalities to a 1% chance of 100,000 fatalities. Risk proneness is identical to risk equity. It means preferring the 1% risk of 100,000 fatalities on the premise that the total possible risks to all individuals are shared equally. Finally, risk neutrality is equivalent to indifference between the smaller risk of a larger number of fatalities and the larger risk of a smaller number of fatalities—the decisions are seen as equally valid.

These are the kinds of considerations that influence the decision-making process when, say, a nuclear power plant is proposed for an earthquake-prone area. To be sure, the real world is a lot messier than our academic example, so it is much harder to estimate potential impacts. Nevertheless, natural hazards are (or should be) evaluated with these kinds of trade-offs in mind.

10-3 RISK ASSESSMENT

The principles of estimating the recurrence time of an earthquake occurrence, a volcanic eruption, or some other natural disaster such as a storm surge are similar. We assign a numerical chance of occurrence within a specified time frame or window of concern.

The occurrence of a natural hazard is not the risk. It is the *outcome* of that occurrence that is important. We need, then, to know the probability of the expected outcome occurring within a specified interval of time. Again, we emphasize the time element. If we could live forever, some natural calamity would surely strike us. This would not be a prediction or a probabilistic estimate of occurrence, but rather a statement that if one lasts long enough, some catastrophe is inevitable.

Suppose a natural disaster strikes that triggers an unfortunate chain of events. These events are not guaranteed to take place but they do have a finite probability of occurring. Here is where we create a causal hazard chain. We assign individual probabilities to all its links and then evaluate the chance that the end outcome will actually happen.

Finally, we factor the human element into our risk assessment. Individual and societal risk will depend on the number of individuals *exposed* to the risk and consequently the number of people *sharing* the risk. At this point we can assess the overall acceptability of the risk or hazard.

10-4 GEOTHERMAL POWER PLANT

A geothermal power plant has been proposed for a site atop an old landslide near San Francisco. It is feared that a new landslide will be triggered by renewed seismic activity at any of three nearby active earthquake faults: the Maacama, Healdsburg–Rodgers Creek, and San Andreas faults (Figure 10-2). To assess the risk potential we need to answer these questions:

1. What is the location of the active faults that have the potential of triggering a damaging landslide beneath the site?
2. What is the pattern of occurrence of earthquakes at these faults in terms of magnitude and recurrence period?
3. What level of ground acceleration can be expected at the site given the subsoil conditions?
4. What is the likelihood that a particular level of ground shaking will trigger a landslide?

To assess risk in this situation, we need a numerical estimate of the overall probability of a landslide being triggered by an earthquake at any one of the three faults during the specified design life of the power plant. We will assume a 30-year design life. (A prespecified value like this can be changed, but if it is, our assessment of risk will also change.)

We formulate the problem as follows. The probability, P, that a landslide will occur is given by the expression

$$P = 1 - \text{probability of no slide}$$

$$= 1 - (1 - P_{SA})(1 - P_{MA})(1 - P_{HR})$$

where the subscripted probabilities represent the probabilities of each of the individual faults triggering a landslide. To assess these individual probabilities, we need to know the maximum-magnitude earthquake that can occur at each fault within our specified design life and also the level of ground acceleration and shaking that will take place at the geothermal plant site as a result of these maximum earthquakes. Finally, we need to know the probability that the estimated level of shaking will start a landslide beneath the power plant.

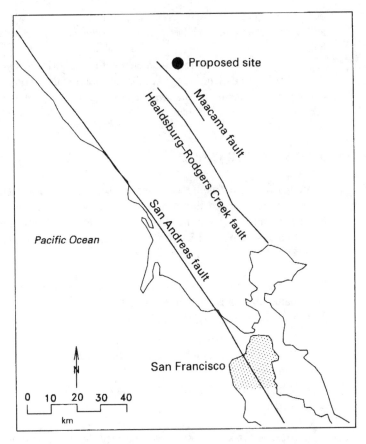

Figure 10-2   Map showing the position of a proposed geothermal plant relative to potentially dangerous earthquake faults in the vicinity of San Francisco, California.

First we determine that the maximum-magnitude earthquake that could occur on the Maacama fault is 6.5. The figures for the Healdsburg–Rodgers Creek fault and the San Andreas fault are $M = 7.0$ and $M = 8.5$, respectively. These numbers are derived from studies of the historical record and empirical investigations that correlate fault length with earthquake magnitude. Briefly, the larger the fault rupture length, the greater the size of the resultant earthquake. Figure 10-3 is a graphic exposition of this assessment. Fault rupture length correlates well with the expected magnitude of a causative fault. We can also look upon these values as our design-basis earthquakes. A *design-basis earthquake* is the *maximum-size* event it is assumed could happen. Smaller events are important cumulatively for the assessment of overall risk, but the larger events always dominate risk estimates.

Our next consideration is the expected return time $T(M)$ for these earthquake events. Remembering that earthquake recurrence data along individual fault seg-

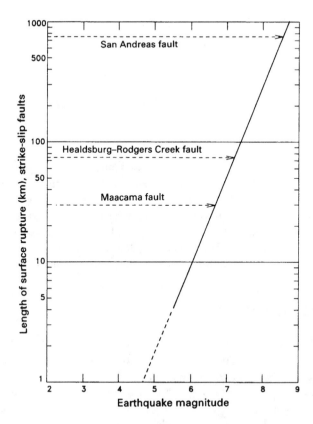

Figure 10-3   An empirical correlation for fault rupture length against earthquake magnitude for strike-slip earthquakes. In strike-slip earthquakes relative displacement on the fault plane is purely horizontal.

ments are expressed in the form $\log N = a - bM$, we determine that the return time for an $M = 8.5$ earthquake at the San Andreas fault is 498 years; for the $M = 6.5$ event at the Maacama fault, it is 4000 years, and for the $M = 7.0$ event at the Healdsburg–Rodgers Creek fault, it is 380 years.

Given these return times, we evaluate the individual seismic risks over our design life of 30 years and conclude that for the San Andreas fault we have [1 − exp (−30/498)] = 0.06 and for the other two faults seismic risks of 0.007 and 0.08. We should remember that our assessment of seismic risk stands or falls on an *accurate* estimate of the average return time for critical-magnitude events on the particular faults.

Our next step is to make a geotechnical assessment of the level of ground acceleration and shaking that would occur at the power plant if the maximum earthquakes occur and then estimate the probability of whether these maximum events would indeed trigger a landslide. This is an empirical geologic judgment, but let us assume that probabilities of 16%, 20%, and 3% are assigned for landslide-triggering events initiated on the San Andreas, Maacama, and Healdsburg–Rodgers Creek faults, respectively.

The individual slide probabilities during our 30-year-lifetime period are

$$P_{SA} = 0.06 \times 0.16 = 0.01$$

$$P_{MA} = 0.007 \times 0.20 = 0.001$$

$$P_{HR} = 0.08 \times 0.03 = 0.002$$

These numbers give the following overall probability of a slide:

$$P = 1 - [(1 - 0.01)(1 - 0.001)(1 - 0.002)] = 0.01$$

What we have done here is to blend the concept of seismic risk—the probability of a seismic event appearing within our specified design lifetime—with the overall probability that the events will take place. Is 0.01 (1%) significant in a 30-year-lifetime? What happens if the design lifetime is extended? What happens if earthquakes occur more frequently than estimated? This example provides a quantitative answer, but what does it mean for prudent decision making? Before we attempt to answer these questions, we will explore another more complicated accident scenario.

10-5 LIQUEFIED NATURAL GAS FACILITY

A number of years ago it was proposed to install a liquefied natural gas (LNG) unloading and docking facility at Little Cojo Bay in Santa Barbara County, California (Figure 10-4). This facility was to accept Indonesian liquefied natural gas that would be shipped via tankers to California at cryogenic temperatures (−259°F), unloaded via cryogenic piping into storage tanks, and later revaporized and piped to consumers. The question was what would happen if an accident, natural or human-made, happened.

Many experts debated the potential effects of various natural hazards—earth-

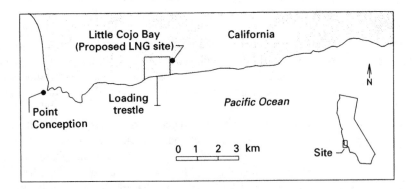

Figure 10-4 The proposed location of a liquefied natural gas unloading and docking facility on the California coastline.

quakes, tsunami, soil creep, landsliding, flooding, erosion, and soil liquefaction—on the proposed site and explored some dangerous natural hazard accident sequences. We will examine one hypothetical possibility.

Point Conception on the California coast was proposed as the site for accepting liquefied natural gas. Suppose a tanker runs aground from a tsunami and its cargo hold is punctured. Remember, the tanker is transporting liquefied natural gas, not oil. Two outcomes are possible. One is that a vapor cloud forms, and the other is that the gas immediately ignites, producing a rapidly spreading pool fire. Either of these results has a finite probability of occurrence. We will create accident scenarios for each of these outcomes.

Our first step is to create a causal hazard chain. Once we have formulated this accident chain of events with their associated probabilities, we can estimate the expected number of human fatalities, which allows us to quantify the public risk. Only then can we make a reasoned evaluation of potential societal harm.

A hypothetical risk analysis scenario is shown in Figure 10-5. The ultimate outcome depends on a number of sequential occurrences culminating in where people are at the time of the accident, a quantification of the risk, and, finally, an evaluation of its acceptability.

First, we estimate the probability that a tsunami will strike at the loading trestle. The occurrence of such an event here is so rare that we will assume an annual probability of one in a million, or 1×10^{-6}. (The daily probability would be much smaller: 2.7×10^{-7} or (0.00000027.) Next we determine the probabilities of the two possible outcomes of the accident: the formation of a vapor cloud versus the immediate creation of a pool fire.

Let us assume there is a 1% chance that a vapor cloud will form and drift shoreward. Our next step then, is to assess the probability of an onshore wind and the probability that a careless workman, visitor, or tourist will accidentally ignite the vapor cloud—for example, by lighting a cigarette, a stove, or a lantern or by starting an automobile.

To calculate the probability of this particular accident sequence—Accident Scenario A—taking place, we will assume a 50% probability of an onshore wind and a 10% likelihood that accidental ignition will occur. The probability of Accident Scenario A, $P(A)$, is

$$P(A) = (2.7 \times 10^{-7})(0.01)(0.50)(0.10)$$

$$= 1.35 \times 10^{-10}$$

The expected number of fatalities is calculated in the following manner. The probability, $P(x)$, of having x number of fatalities is the product of x fatalities if Accident Scenario A occurs multiplied by the probability, $P(A)$, that the accident sequence will occur (see Appendix A). In other words, the probability $P(x)$ of having x fatalities is

$$P(x) = P\left(\frac{x}{A}\right)P(A)$$

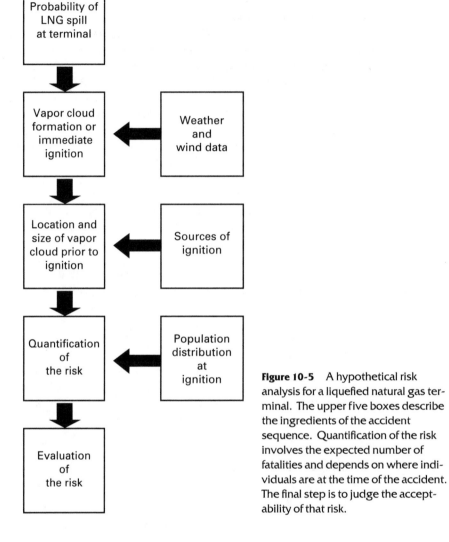

Figure 10-5 A hypothetical risk analysis for a liquefied natural gas terminal. The upper five boxes describe the ingredients of the accident sequence. Quantification of the risk involves the expected number of fatalities and depends on where individuals are at the time of the accident. The final step is to judge the acceptability of that risk.

and the expected number of fatalities due to Accident Scenario A is

$$x \cdot P\left(\frac{x}{A}\right)$$

The *societal risk*, $F(A)$, is

$$F(A) = P(A) \cdot \text{expected number of fatalities}$$

To assess the overall risk, we need to sum the possibilities of all individual accident chains with their individual expected fatalities and create an annual risk or expectation. To do this, we need to introduce some additional definitions. Individual risk, R, is specified as the societal risk, F, divided by the number, N, of people exposed to the risk:

$$R = \frac{F}{N}$$

We can additionally specify a group risk, R_G, where R_G is the probability that a member in a group, G, of individuals (surfers, skiers, individuals with tattoos, whatever) would be a statistic:

$$R_G = \frac{F_G}{N_G}$$

To reiterate, *societal risk* is the product of the probability of the accident sequence taking place times the number of expected fatalities. *Individual risk* is societal risk divided by the total number of people exposed to the hazard. It should be emphasized that the end consequence depends on the probability of a particular accident scenario taking place and our estimate of the expected number of fatalities.

We determined that the probability of Accident Scenario A occurring was 1.35×10^{-10}. Suppose 400 workers are the expected number of fatalities. In this case, the societal risk would be

$$F = P(A)(400) = 1.35 \times 10^{-10} \times 400 = 5.4 \times 10^{-8}$$

Observe that societal risk is directly proportional to the expected number of people affected and the probability that on any given day the sequence of Accident Scenario A will indeed take place.

Now let us consider Accident Scenario B: A liquefied natural gas pool occurs with a high probability (say 0.99) of immediate ignition. In this instance, our societal risk would be equal to

$$F = (2.7 \times 10^{-7})(0.99)(400)$$

or

$$F \sim 1.1 \times 10^{-4}$$

It is clear that our societal risk depends strongly on the number of people sharing the risk at the time of the accident. Suppose the calamity occurs during the heavy tourist season. In this case, the risk per person per year is greatly reduced, because we are spreading the risk. If 4000 tourists are present rather than 400 workers in Accident Scenario B, the *risk per individual* on an *annualized* basis will be the societal risk divided by the 4000 people exposed:

$$2.8 \times 10^{-8} \quad \text{or} \quad 0.000000028$$

The result looks like a very small number, but is it? We shall explore this question in the following section.

10-6 RISK ACCEPTABILITY

Acceptable risk from earthquakes is defined in California by seismic regulations in terms of building damage: Minor earthquakes should not produce damage, moderate earthquakes should not produce any significant structural damage, and major or severe earthquakes should not produce a total failure such as building collapse. It is necessary to correlate these definitions of "minor" and "major" with an expected intensity level of shaking. The true assessment of risk, however, must be balanced against other everyday risky human activities. Economic loss is another prime consideration.

We want to decrease the individual risk posed by a hazard—that is, we want to reduce the likelihood of any deaths. Table 10-1 lists some of the risks encountered in daily life. These risks are estimations that increase the chance of death in any year by 1 part in 1 million (1×10^{-6}). Many of us are exposed to one or more of these risks everyday, but they customarily provoke little concern. Accidents, on average,

Table 10-1   Some Everyday Risk Activities

Activity	Cause of Death
Smoking 1.4 cigarettes	Cancer, heart disease
Drinking 0.5 liter of wine	Cirrhosis of the liver
Spending 1 hour in a coal mine	Black lung disease
Spending 3 hours in a coal mine	Accident
Living 2 days in New York or Boston	Air pollution
Traveling 6 minutes by canoe	Accident
Traveling 10 miles by bicycle	Accident
Traveling 150 miles by car	Accident
Flying 1000 miles by jet	Accident
Flying 6000 miles by jet	Cancer caused by cosmic radiation
Living 2 months in Denver on vacation from New York	Cancer caused by cosmic radiation
Living 2 months in average stone or brick building	Cancer caused by natural radioactivity
One chest X-ray taken in a good hospital	Cancer caused by radiation
Living 2 months with a cigarette smoker	Cancer, heart disease
Eating 40 tablespoons of peanut butter	Liver cancer caused by aflatoxin B
Drinking Miami drinking water for 1 year	Cancer caused by chloroform
Drinking 30 12-oz cans of diet soda	Cancer caused by saccharin
Living 5 years at site boundary of a typical nuclear power plant in the open	Cancer caused by radiation
Eating 100 charcoal broiled steaks	Cancer from benzopyrene

SOURCE: Wilson, 1979.

NOTE: The risks shown increase the chance of death by 1 death in 1 million individuals exposed to the risk.

shorten a life span by about 30 years. So the individual risk of one in a million from an accident therefore shortens life by 30×10^{-6} years or 15 minutes.

To understand a risk of one in a million, consider cigarette smoking. In the United States it has been estimated that 627 billion cigarettes are smoked yearly. Not everyone smokes, but this number of cigarettes averages out to 3000 per person or $\frac{1}{2}$ pack a day per individual. Approximately 15% of all Americans die from lung cancer (the percentage is larger if only smokers are considered). The individual lifetime risk of getting lung cancer, then, is 0.15 divided by a 70-year-life expectancy, giving an individual yearly risk of 0.002. The risk per smoker is 0.002/3000 or about one in a million. The time it takes to smoke one cigarette, then shortens life expectancy by that same period of time.

An alternative evaluation of the balancing of risk is shown in Table 10-2. Notice that the risk of death from natural hazards is considerably lower than the risks presented by certain everyday activities. Notice also that people's risks of dying from a natural hazard vary widely depending on what country or regions they live in.

Table 10-2 sheds some light on people's different attitudes toward voluntary and involuntary activities. Voluntary activities are those things, like playing sports and driving a car, that individuals do out of choice. They adjust to the levels of risk of these activities and gauge their acceptability. A typical level of acceptability is in the range of 1 death per 10,000 to 20,000 persons exposed to the risk per year. Involuntary risks are those risks over which individuals have no control, such as rail-

Table 10-2   Probability of Dying in any One Year
from Various Causes

Smoking 10 cigarettes a day	One in 200
All natural causes, age 40	One in 850
Any kind of violence or poisoning	One in 3300
Influenza	One in 5000
Accident on the road (driving in Europe)	One in 8000
Leukemia	One in 12,500
Earthquake, living in Iran	One in 23,000
Playing field sports	One in 25,000
Accident at home	One in 26,000
Accident at work	One in 43,500
Floods, living in Bangladesh	One in 50,000
Radiation, working in radiation industry	One in 57,000
Homicide living in Europe	One in 100,000
Floods, living in northern China	One in 100,000
Accident on railway (traveling in Europe)	One in 500,000
Earthquake, living in California	One in 2,000,000
Hit by lightning	One in 10,000,000
Wind storm, living in northern Europe	One in 10,000,000

SOURCE: Coburn and Spence, 1992.

road and airplane crashes, gas explosions, and natural hazards. They may have a small advantage by selecting which airline to travel on and when to travel, and certain natural hazards pose a greater risk than others. Nevertheless, people are willing to accept voluntary risks that are 100 to 1000 times greater than involuntary risks.

The observation that individuals are willing to accept voluntary risk levels much higher than those expected from natural disasters has an impact on post disaster decisions for reconstruction. There is a strong tendency to resettle in the same area struck by the disaster. The explanations are plausible. One factor is the strong human desire to return. Another factor is that reconstruction offers psychological benefits. The human tendency or rationalization—that perhaps history will not repeat itself—is pervasive. A final consideration is that humans are continually hopeful that modern technology will prevent destruction in future natural disasters.

So what should be the criterion for risk acceptability? Risk acceptability balances the annual probability of death per person exposed against the expected social benefit. It is often argued that where human life is concerned, it is best to err on the conservative side. The acceptable risk level is typically specified as one in ten million (1×10^{-7})—that is, 1 death per 10 million persons.

It is easy to specify a level of risk as low as 1 in 10 million but can we afford such a stringent standard? It is impossible to ignore the cost question when specifying an acceptable level of mitigation or protection. We will always find it necessary to balance caveats when evaluating the impact of a potential natural hazard. If, say, we propose to build a controversial power facility in an area susceptible to natural hazards, we have to consider environmental and socio-economic factors, public safety and health, and individual attitudes. Uncertainties, value tradeoffs, and individual perceptions of the acceptability of the risk and the desirability of equity are the ingredients that shape such decision making.

The world's population is expected to double over the next century, and the bulk of this increase will be in the developing countries, which are already most susceptible to natural disasters. One of the greatest challenges at the end of the twentieth century is to design and implement policies that will offset the inevitable increases in disaster potential. Only our understanding and systematic mitigation of natural hazards can bequeath to generations a safer Planet Earth.

REVIEW

1. What is a causal hazard chain and what are its shortcomings?
2. A LNG tanker is anchored off the California coastline. An earthquake in Japan generates a tsunami that subsequently slams into the anchored tanker, producing a spill. Assume that the annual probability of this disaster is 0.01% and that there is a 99% probability of no immediate ignition, so a vapor cloud forms. There is a 90% probability of an onshore wind, and you are one of 4000 people on the beach.

There is a 50% probability that the first individual who lights a cigarette will ignite the cloud and 2000 fatalities will result. What is the annual societal risk?

3. What is your individual risk in the above accident scenario?

4. Is the risk acceptable? Explain.

5. You are one of 20 water-skiers being towed by your own speedboat with its own driver on a reclaimed lake in the Sierras that is sided at the upper end by a very unstable pile of mine tailings that could easily be triggered by an earthquake, which would fill the lake and bury you. You and your friends water-ski daily all year, rain or shine, and are the expected fatalities. Eighty perennial campers, skinny dippers, and other associated types on the shoreline are also exposed to the risk. If your individual annual risk is 0.01, what is the annual probability that the fatal landslide will be triggered?

6. Suppose that a large earthquake has a return time of 24 hours and has a 40% probability of collapsing the building that you are in. You have a 30% probability of surviving the collapse. What is the overall risk of your not surviving the next 3 hours?

7. Is the annual probability of being a fatality in an earthquake disaster greater in California or Iran? Explain.

8. What is the difference between group risk and individual risk?

9. You are a student residing in a dormitory who is paranoid about earthquakes. You know that an $M = 7$ earthquake is anticipated within a year on a nearby fault. You are told that for your area the earthquake magnitude-frequency relation is

$$\log_{10}N = 4.23 - 0.815M$$

Under the following list of provisos, what is your annual risk of dying in your dormitory?

- You have opted for year-round residence.
- You spend half of your day in your dormitory.
- The probability of your dormitory collapsing from an $M = 7$ quake is 50%.
- There is a 30% chance of surviving the collapse.
- There is a 1% chance of a subsequent nonfatal fire.

10. In the landslide-triggering problem discussed in the text, the San Andreas fault was determined to have a seismic risk of 6% within 30 years. If this risk is actually 100%, what is the overall probability that a landslide would be triggered?

11. Under what conditions could the above overall probability be ~ 100%?

Basic Probability Theory

Probability is that branch of mathematics concerned with the chances that a given event will occur. The probability of an event occurring, $P(E)$, is defined as the ratio of the favorable outcomes, m, to the total number of possible events, n:

$$P(E) = \frac{m}{n}$$

Probability values can range from 0, meaning the event can never happen, to 1, signifying a sure outcome. Probabilities are often expressed as a percentage ranging from 0% to 100%. If P is the probability of an event taking place, then $1-P$ is the probability that it will not occur.

Consider a die, with values from 1 to 6. The probability of throwing an even number is 3/6 or 1/2. The probability of throwing either a 4 or a 5 is 2/6 or 1/3. People often confuse *probability* with *odds*. If P is the probability that A will occur, then $1-P$ is the probability that it will not occur. The odds that A will occur is equal to $P/(1-P)$. In the gambling community odds has a slightly different meaning. When odds of 5 to 2 are offered in favor of event A occurring, the bookie is offering \$5 that A will occur for every \$2 you bet that A will not occur. The ratio of his bet to yours—5 to 2 in this case—is called the *betting odds*. The probability of obtaining a 7 or an 11 on a dice throw is 8/36. The odds, therefore, are 8/28 or 2/7. If you are betting in favor of a 7 or an 11, then for every \$2 you wager, your opponent should offer \$7.

If in a trial with n possible results the event F occurs k times, its probability $P(F)$ is k/n. Now if for m of these k events F further satisfies a condition such that event E

also occurs, its probability $P(E/F)$, is equal to m/k. Therefore, m/n also represents $P(EF)$ since EF denotes an event that satisfies both the conditions for F and E to occur. That is, $P(EF) = P(F)P(E/F) = km/nk = m/n$. As an example, consider that in the roll of two dice, 36 outcomes are possible. What is the probability that the sum of the dice is divisible by 2 and also divisible by 3? In this case $P(23) = P(2)P(3/2)$, giving us $(18/36) \times (6/18)$ or 1/6. This is an example of the *multiplicative law for probabilities*.

Another probability concept is that of *independent events*. Two events, E and F, are considered to be independent of each other if the occurrence or nonoccurrence of one event has no influence on the outcome of the other. In throwing two dice, the score obtained with one die does not depend on the value obtained with the other. If E denotes a score of 4 on one die and the dice are distinguished by the indices 1 and 2, $P(E_1) = 1/6 = P(E_2)$. However, $P(E_2/E_1) = 1/6$ also. The probability of throwing a pair of 4's is $P(E_1E_2) = (1/6) \times (1/6) = 1/36$. This leads us to the multiplication law for the probability of independent events:

$$P(E_1E_2) = P(E_1)P(E_2)$$

B

Binomial and Poisson Distribution

We wish to develop a formula for the probability of achieving exactly k successes out of n independent trials. Let 1 denote a success and 0 a failure. We want to find $P(k$ 1's in n trials). Suppose we assume $k = 3$ and $n = 5$. One way to obtain 3 successes is 11100. The probability of this event occurring is $P(11100)$, which by the independence of the trials is $P(1)P(1)P(1)P(0)P(0)$ or $pppqq = p^3q^2 = p^3q^{5-3}$, where p is the probability of success and q that of a failure. Now 10011 is also a success, so

$$P(10011) = pqqpp = p^3q^{5-3}$$

Observe that $P(11100) = P(10011)$. Therefore, $P(3$ 1's in 5 trials$) = P(11100$ or 10011 or . . .), where the dots represent all the other ways of achieving 3 successes in 5 trials. All these events are mutually exclusive. Therefore,

$$P(3 \text{ 1's in 5 trials}) = P(11100) + P(10011) + P(11010) + \ldots + p^3q^{5-3} + p^3q^{5-3} + p^3q^{5-3} + \ldots$$

$$= rp^3q^{5-3}$$

where r is the *total* number of ways of getting 3 successes in 5 trials. We need to find r.

There are 5 positions available in which to place the first 1, $5-1$, for the second, and $5-2$ for the last position. In other words, there are $5(5-1)(5-2)$ *different positional* arrangements. We are going to write this in shorthand as $5!/2!$, where $5! = 5 \times 4 \times 3 \times 2 \times 1$ and $2! = 2 \times 1$; $5!/2! = 5 \times 4 \times 3 = 5(5-1)(5-2)$.

In addition, for any *fixed* position of three 1's, the 1's could be permuted among themselves—there are 3! (6) such permutations. All such permutations are indistinguishable since the objects are *identical*. Therefore, there are $r = (5!)/(2! \times 3!)$ or 10 unique arrangements among the places, giving us the formula for calculating the probability of achieving k successes in n trials:

$$(n!\, p^k q^{n-k})/k!(n-k)!$$

This probability is called the *binomial distribution.*

An important approximation to the binomial distribution takes place when the probability of success, p, is small and the number of trials, n, is large. If we define λ as the expected *number* of occurrences,

$$\lambda = (\text{number of trials}) \times (\text{probability of event}) = np$$

It can be mathematically demonstrated that the probability of k successes in a *given unit of time* Δt is

$$P(X = k) = [\exp(-\lambda\, \Delta t)](\lambda\, \Delta t)^k/k!$$

This probability relation is called the *Poisson distribution.*

Since this event process has many applications, such as earthquake sequences as a function of time, it is sometimes called the *model of catastrophic events.* Statistically, large earthquakes are rare events. The concept of independent rare events means that their distribution in time approaches the behavior predicted by the Poisson distribution.

Since the expected value is $E(X) = \lambda\, \Delta t$,

$$\lambda = \frac{E(X)}{\Delta t}$$

λ is called the average rate of failure. $T = 1/\lambda$ is then called the "time to failure."

An important facet of the Poisson process is the homogeneity of time. The probability of k events occurring in one interval of time is *identical* to the probability of k events occurring in any other nonoverlapping time interval of the same length. Since the *expected* value of k events occurring in a time interval is $\lambda\, \Delta t$ we call $T = 1/\lambda$ the "time to failure" or the recurrence time of the event.

APPENDIX

C

Seismic Moment

Seismic moment is a measure of earthquake size related to the leverage of the forces applied across an area of fault slip. It expresses the mathematical equivalence between forces and the size of a shear dislocation.

Imagine that a small orange is centered on the focus or hypocenter of an earthquake. Suppose that the orange is sliced in a nearly horizontal plane that schematically represents the plane of slip along a fault or failure plane in the earth. If the plane labeled "Fault Plane" in Figure C-1 is the plane of rupture, then another plane in space passing through the earthquake focus is the *auxiliary plane*. We call the fault slip plane a *plane of shear dislocation*. A shear dislocation is a surface possessing a discontinuity in displacement on opposite sides of a fault plane. The dislocation surface is abstractly considered to be welded together so that there is no discontinuity in stress across the surface, although there are strains (the geometrical change in length or volume divided by the original value) because of the discontinuous displacement across the zone.

In Figure C-2 the horizontal line represents the surface trace of a vertical fault plane and the vertical line represents the projection of the auxiliary plane. Can a system of external forces be created that generates the same strains even if the dislocation is removed? The answer is yes, and the orthogonal moments (force $\times$ lever arm) of these forces is $M_o = \mu DA$, which is the product of the rigidity of the rock, μ, times the amount of slip or offset, D, times the fault area (length $\times$ width of rupture plane), A. M_o is defined as the *seismic moment*.

Another important concept is the *radiation pattern of earthquake waves*.

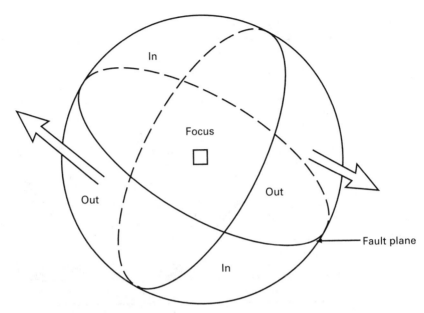

Figure C-1   An imaginary sphere centered on the focus of an earthquake to illustrate the concept of a fault-slip plane and its imaginary orthogonal, auxiliary plane.

Depending on which spatial quadrant an observer is situated in relative to the earthquake's focus, the sense of motion felt—either an outward push or an inward jerk—will give a clue as to the orientation of the fault plane and its imaginary conjugate plane in space. This idea is shown in Figures C-3 and C-4. If the fault slip plane is purely horizontal or *strike slip*, the resultant pattern in our hypothetical orange will project onto a horizontal plane as a quadrantal pattern of outward pushes and inward pulls from the focus (in Figure C-3). In the type of faulting known as *thrust* or *normal faulting*, where relative motion is upward and downward along the fault plane, the pattern of pushes and inward pulls relative to the focal point will be completely different (Figure C-3). Mixtures of slip motion along a fault plane are possible and can be resolved if the spatial and azimuthal coverage of the pattern of pushes and inward pulls can be discerned.

The method involves plotting the pattern of the polarity (push or pull) of the first motions radiated by the earthquake focus on to a horizontal plane. A hypothetical sphere is again centered about the focus (Figure C-5). A pole of projection at the top of the sphere is selected from which a plane such as *AOBC* in space can be projected downward onto a horizontal plane and its orientation in space determined. The procedure involves the positioning in space of two orthogonal planes—the fault plane and its auxiliary plane—which correctly separate into spatial quadrants the pattern of pushes and pulls radiated by the earthquake source.

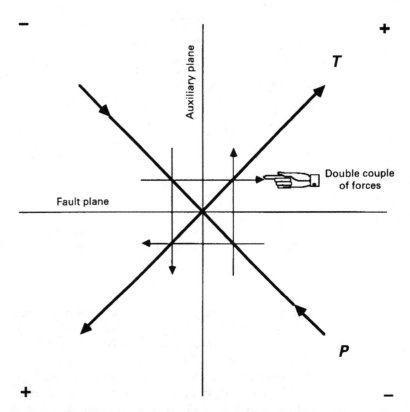

Figure C-2   The surface expression for a vertical fault and its auxiliary plane. The horizontal line represents the intersection of the fault plane with the earth's surface. Relative motion is right lateral in that the opposite side of the designated fault plane moved to the right. The result is to produce a push (+) in the upper-right-hand and lower-left-hand quadrants. The pattern of pushes and pulls is produced by the double couple of forces shown, or by a system of forces labeled P and T that bisect the quadrants of pushes and pulls.

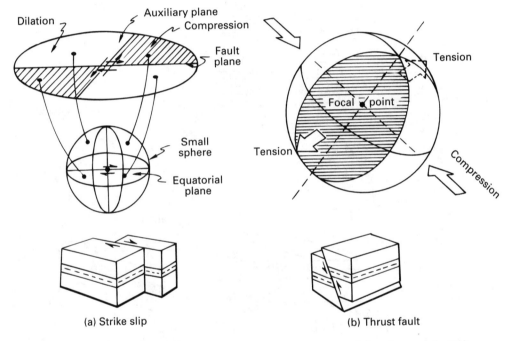

(a) Strike slip   (b) Thrust fault

Figure C-3   The radiation pattern of the initial sense of motions produced by (a) strike-slip and (b) thrust (reverse) faulting.

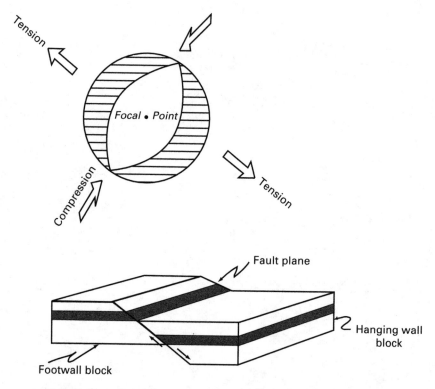

Figure C-4   A diagram showing normal faulting and the pattern of the first motions projected on an infinitesimal focal sphere. In normal faulting the hanging wall block moves downward on the fault plane relative to the foot-wall block.

Pole of projection

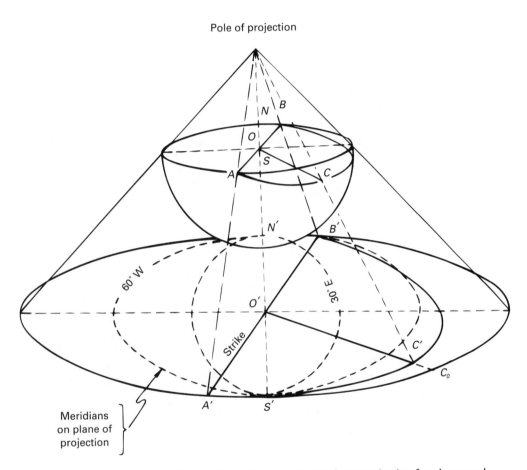

Meridians
on plane of
projection

Figure C-5   The principle of a lower-hemisphere projection for planes and
lines in space.

APPENDIX

D

Volcanic and Seismic Risk Analyses

Some facts can be explained in several ways even when there is actually only one correct explanation.

—*Lucretius, De rerum natura*
[On the nature of things]

We will introduce a probabilistic estimate to evaluate risk of volcanic and seismic activity. Bear in mind that any such estimate does not necessarily predict what will happen, but rather what *would* happen if our assumptions are accurate and our postulates correct. We may never know the correct explanation of why a volcano erupts or an earthquake occurs, but we can *evaluate* the risks involved given some plausible assumptions.

D-1 POISSONIAN MODELS OF VOLCANIC AND SEISMIC ACTIVITY

Many volcanoes appear to erupt at irregular intervals. We here define the interval between eruptions as the repose time, t. The time when the repose period ends is defined as the eruption time. A certain order appears when the number of times that a particular volcano has a repose period greater than or equal to t is counted.

Consider the apparently irregular eruption history of the volcano Popocatépetl in Mexico shown in Table D-1. Starting with these data, we create a second table where we tabulate the frequency or number of times, N, for which the repose time is

Table D-1   Eruption History of Popocatépetl

Eruption Times	Repose Times
1519	
	11
1530	
	9
1539	
	3
1542	
	6
1548	
	23
1571	
	21
1592	
	50
1642	
	22–25
1664–1667	
	30–33
1697	
	23
1720	
	82–84
1802–1804	
	118
1920	

greater than or equal to t (Table D-2). If we next plot the natural logarithm of N versus t, the data can be approximately fit by a linear relation of the form $\ln N = \alpha - \beta t$.

We define the slope of this line, β, as the age-specific eruption rate of the volcano. For many volcanoes, the age-specific eruption rate is independent of time. This is equivalent to stating that the chance of an eruption is independent of time and has no memory. We call these volcanoes "simple Poissonian volcanoes" and make

Table D-2   Popocatépetl Volcano

Number of Reposes ≥ t Years	
N	t
12	3
11	6
10	9
9	11
8	21
7	22
6	23
4	30
3	50
2	82
1	118

use of the statistics of a Poisson process. That is, we write that the probability of n eruptions, P_n, occurring within a time interval Δt is given by the expression

$$P_n = \frac{(\beta \Delta t)^n}{n!} \exp(-\beta \Delta t)$$

β is the age-specific eruption rate of the volcano and $n!$ is shorthand for n factorial (see Appendix B).

Suppose a volcano has an age-specific eruption rate $\beta = 0.013/month$. This volcano would have the following probabilities for zero, one, and two eruptions within 20 months:

$$P_0 = 0.77; P_1 = 0.20; P_2 = 0.03$$

Observe that there is a high probability that no eruptions will occur within the next 20 months and a 20% probability that at least one eruption will occur within this time frame. It is apparent that the eruption probability is dependent on both the time interval, Δt, and the eruption rate. For a fixed Δt, a slow eruption rate predicts a high probability that no eruption will occur in the immediate future. On the other hand, a higher eruption rate indicates low probabilities of long repose intervals for the volcano.

We now turn to earthquake activity patterns. We are assuming that the individual seismic events constitute a Poisson process—in essence, that the occurrence of individual earthquakes of a specified magnitude is independent and identically distributed. Then the probability of having n earthquakes with a magnitude exceeding critical magnitude M during the time interval ΔT is equal to

$$P_n = \frac{\exp(-\lambda \Delta T)(\lambda \Delta T)^n}{n!}$$

where λ is defined as the average rate that a magnitude M earthquake will be exceeded in a given spatial volume over a specified time interval.

If we state that no earthquakes will occur ($n = 0$) within the time interval ΔT, the probability that our earthquake will *not* occur during our specified interval ΔT is simply

$$\exp(-\lambda \Delta T)$$

This equation is equivalent to stating that the probability that our event will occur within the time interval ΔT is

$$1 - \exp(-\lambda \Delta T)$$

We showed in the text that earthquake recurrence data can be conveniently expressed in the form

$$\log_{10} N = a - bM \quad \text{or} \quad N = 10^{(a-bM)}$$

where N is the number of earthquakes of magnitude M or greater that occur in a region per unit of time, typically taken to be 1 year. That is, we now assume that $N = \lambda$, the rate of the process.

Using the concept of an average recurrence time, we observe that the probability of our selected event occurring or being exceeded in our time interval, ΔT, is given by

$$1 - \exp\{-10^{(a-bM)}\Delta T\}$$

leading to our final statement that the probability is

$$1 - \exp\{-\Delta T/T(M)\}$$

where we have made use of our previous observation that $N = 1/T(M)$. This relation forms the basis for the evaluation of seismic risk.

D-2 VOLCANIC RISK ASSESSMENT

In most cases, it is not possible to change the outcome of a volcanic eruption, nor is it possible to put a monetary value on the loss of human life. It is possible, however, to estimate the other costs due to an eruption, and in doing so, to provide a rational basis for policy planners and local populations to assess the risks of development in the area. As we said at the outset of this appendix, whenever we are dealing with natural phenomena, there will be some uncertainty about the outcome. The best we can do is to address outcomes and consequences in a probabilistic fashion and then decide upon a rational course.

Let us take a hypothetical example to illustrate this procedure. Suppose the owner of a large resort complex on a volcanic island is uncertain whether to continue the present rate of development and maintenance of the property or discontinue development and maintenance altogether and assign no more funds to the complex. We will call these two monetary incomes A_1 and A_2, and for simplicity's sake we will restrict the volcano to two conditions: θ_1 for in eruption and θ_2 for not in eruption. We will also consider only two possible outcomes: θ_{11} for the destruction of all economic value and θ_{12} for the partial destruction—say, 1/5—of the economic value. Obviously, we could consider other outcomes, but this will do for a brief illustration of how the procedure works.

We define V_0 as the economic value of the complex (land and improvements) and I_0 as the amount of money that would be dedicated to continuing development, improvements, and maintenance. Clearly, the expected outcomes $E(A_1)$ and $E(A_2)$ are dependent upon our assessments of the probabilities of θ_1, θ_2, θ_{11}, and θ_{12} occurring. To evaluate the possible outcomes, we construct a decision tree (Figure D-1)—a diagram in which the branches represent choices with associated probabilities.

In our case, the expected outcome can be written as

$$E(A_1) = P(\theta_2) \times 0 + P(\theta_1)\{P(\theta_{11}) \times [-(V_0+I_0)] + P(\theta_{12}) \times [-1/5(V_0+I_0)]\}$$

$$E(A_2) = P(\theta_2) \times (-I_0) + P(\theta_1)\{P(\theta_{11}) \times (-V_0) + P(\theta_{12}) \times [-1/5(V_0)]\}$$

For simplicity, let us assume that $P(\theta_{11}) = P(\theta_{12}) = 0.5$ and that $V_0 = 5000$ and $I_0 = 2500$. The results are shown below for various probabilities of eruption and noneruption. (It also should be realized that $P(\theta_1) + P(\theta_2) = 1$.)

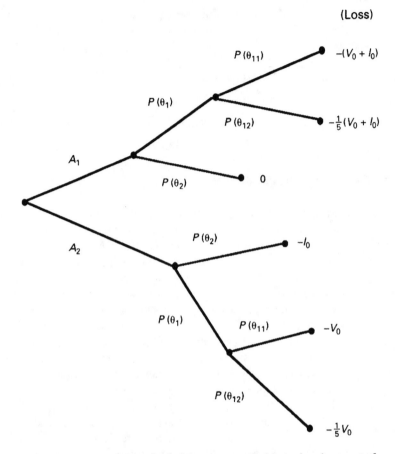

(Loss)

$P(\theta_{11})$ $-(V_0 + I_0)$

$P(\theta_1)$

$P(\theta_{12})$ $-\frac{1}{5}(V_0 + I_0)$

A_1

$P(\theta_2)$ 0

$P(\theta_2)$ $-I_0$

A_2

$P(\theta_1)$ $P(\theta_{11})$ $-V_0$

$P(\theta_{12})$ $-\frac{1}{5}V_0$

Figure D-1   An example of a decision tree applicable to development of a resort complex on a hypothetical volcanic island.

$P(\theta_1)$	$P(\theta_2)$	$E(A_2)$	$E(A_1)$
0.9	0.1	−2950	−4050
0.8	0.2	−2900	−3600
0.7	0.3	−2850	−3150
0.6	0.4	−2800	−2700
0.5	0.5	−2750	−2250
0.1	0.9	−2550	−450

If there is a high probability that the volcano will erupt, the owner stands to lose more money by continuing development. With a 60% chance of the volcano erupting,

the expected monetary consequences of continuing development are about comparable to the consequences of stopping development. Other scenarios could be constructed that would include other natural hazards, but this example is sufficient to
demonstrate the decision-making process in these situations.

D-3 SEISMIC RISK ASSESSMENT

Earthquake recurrence data also satisfy an exponential distribution as a Poisson
process. We define earthquake risk as the probability that a critical earthquake of
magnitude $\geq M$ will take place within a specified interval of time ΔT.

$$P(\text{earthquake} \geq M \text{ in } \Delta T \text{ years}) = 1 - \exp\left[\frac{-\Delta T}{T(M)}\right]$$

This is equivalent to stating that the probability of an earthquake with a magnitude less
than M is

$$\exp\left[\frac{-\Delta T}{T(M)}\right]$$

where $T(M)$ is the mean return period for our designated-magnitude earthquake.

For example, if the average return period for our critical event is 360 years and
our exposure time is 30 years, we have

$$P = 1 - \exp\left[\frac{-30}{360}\right] = 0.08$$

meaning there is a 8% probability that our critical event will occur within the selected
time interval.

If we lengthen the exposure time from 30 to 100 years, $P = 0.24$, and if the time
is lengthened to the mean return period for our specified-magnitude event, $P = 0.633$.
This tells us there is a *more than even chance* that our critical earthquake will take
place.

As shown earlier, earthquake recurrence data are analyzed using a relation of the
form

$$\log_{10} N = a - bM$$

where N is the number of earthquakes per year with magnitudes greater or equal to M.
Thus the mean return period, T, is given by

$$T(M) = \frac{1}{N(M)} = 10^{bM-a} \text{ years}$$

In central and northern California the cumulative number of earthquakes
expected with a particular magnitude is given by

$$\log N = 4.23 - 0.815M$$

Table D-3   Seismic Risk Data for Central and Northern California (Risk %)

$\geq M$	N	$T(M)$	Day	Week	Month	Year	Decade
4	9.3	0.107	2.5	16	54	100	100
5.5	0.56	1.79	0.15	1.1	4.5	43	100
6.5	0.086	11.63	0.023	0.16	0.71	8.2	57
7	0.033	30.30	0.009	0.064	0.28	3.2	27

When we use the data shown in Table D-3 to assess the seismic risk in the area, we find that the chance that an earthquake of $M = 6.5$ will take place within a decade in central or northern California is better than 50/50.

In California then, we can expect one magnitude 8 shock about every 100 years. This anticipation leads to several conclusions worth contemplating:

1. The risk of an $M = 8$ event occurring in any given year is 1%.
2. The probability of at least one such event occurring in a decade is 10%, and the probability of its occurrence in a 50-year time interval is 39%.
3. The probability that at least one such event will take place in a century is 63.3%.

To further illustrate assessment of seismic risk, we will use another example: the Dead Sea Rift Zone. The Dead Sea Rift Zone extends northward from the northern end of the Red Sea for a distance of more than 800 km. There are several linear fault-controlled topographic depressions along portions of this rift zone. The Dead Sea Rift is a transform plate boundary connecting the Red Sea, where crustal spreading is occurring, northward to a zone of plate convergence. The Arabian plate lies to the east of the rift; on the west, several smaller plates, the largest of which is the Sinai microplate, form a part of the larger African plate (Figure D-2).

Historically, the primary earthquake region in terms of damaging activity is that portion of the Dead Sea Rift that extends from 30.8° N to 33.3° N latitude. Figure D-3 is a plot of the earthquake recurrence relation from 2150 B.C. to 1979, compiled from historical sources and recent instrumental data. These data support the empirical relation

$$\log_{10}N = 3.10 - 0.86M$$

where N is the number of earthquakes per year with a magnitude greater than or equal to M. This relation gives a mean return period, $T(M)$, of

$$T(M) = 10^{\,0.86M-3.10} \text{ years}$$

For example, a return period of 115 years is anticipated for an $M = 6.0$ event and an average return period of 2 years is anticipated for a $M = 4.0$ earthquake.

Suppose we are interested in determining the risk of an $M = 6.0$ earthquake occurrence within the next 50 years. We then have

$$\text{Risk} = 1 - \exp\left[\frac{-50}{115}\right] = 0.33$$

meaning that there is a better than 30% chance that such an event will occur. Note that such risk assessments are dependent on the historical database. Since some historical data may well be missing or never present, we must view these types of probabilistic estimates with a critical eye.

Actually, the key parameter for the evaluation of seismic risk anywhere is ground acceleration. Usually risk maps are prepared to depict the *maximum acceleration with a 10% probability of being exceeded in 50 years*. This is the same as saying that there is a 90% probability that the maximum acceleration will *not* be exceeded in 50 years. Figure D-4 is a map of the United States showing the probability that a spec-

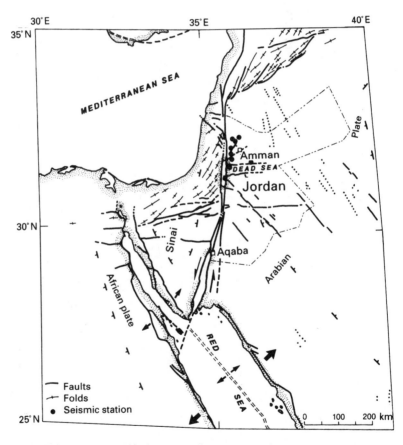

Figure D-2   Geographic location of the Dead Sea Rift Zone. The Rift Zone is the major north-south fault zone on the western side of Jordan. The location of some seismic stations in Jordan are shown by the large solid circles. The smaller solid circles are volcanic domes. (From Kovach and Healy, 1990.)

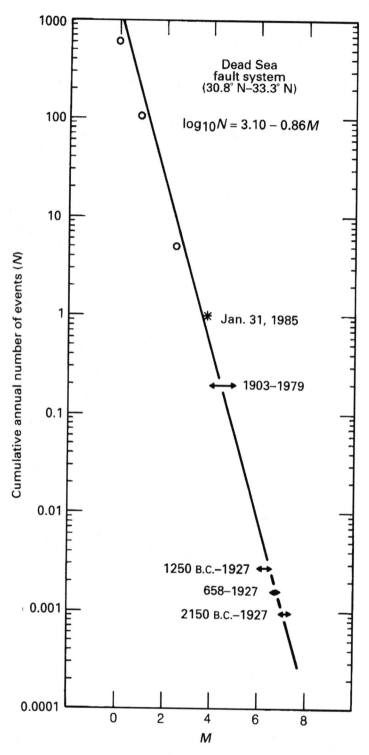

Figure D-3 Earthquake recurrence relation for the Northern Dead Sea Rift Zone. (From Kovach, 1988.)

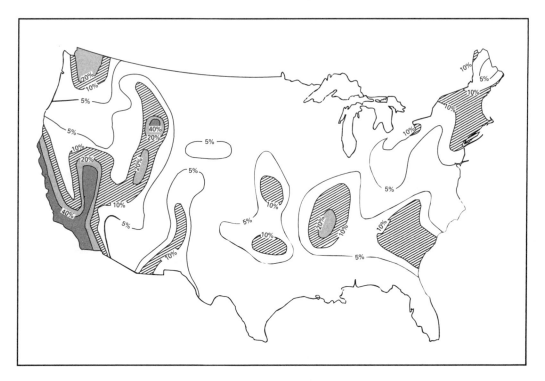

Figure D-4   A seismic risk map for the United States. The numbers indicate the percentage of gravity with a 10% probability of being exceeded in 50 years. This is equivalent to stating there is a 90% probability that ground acceleration will *not* exceed this percentage of gravity in 50 years. Notice that the highest probabilities for damaging ground accelerations are found in California. (Courtesy Applied Technology Council, 1978.)

ified acceleration will be exceeded at least once during the exposure time, ΔT, of 50 years. The map makers used the past distribution of historical seismic events, geologic trends, and reports concerning felt events to assess the size and frequency of earthquakes across the country. The map is deliberately simple and smooth; it does not reflect the concept of microzonation. But it does give a good picture of where dynamic ground accelerations from a future earthquake can be expected in the United States.

In summary, the average return or recurrence time of an earthquake of a given magnitude is determined from a catalog of past historical occurrences. An average return time of 100 years means that we should expect 5 earthquakes within 500 years. Their occurrence is usually irregular, however. Thus one earthquake might take place within 20 years after the previous event, and then be followed by a second occurrence 200 years later. Even though the individual earthquakes occur irregularly in time, an average rate of occurrence of a specified magnitude event can be defined.

Glossary

Aa   The Hawaiian word for *lava* characterized by a rough, clinkery, or jagged surface.

Acceleration   The rate of increase of speed or velocity. Ground accelerations due to earthquake motions are measured relative to 980 cm/sec² or 1 *g*.

Active fault   A *fault* in which movement has occurred in historical time or that now gives evidence of earthquake activity.

Aftershocks   Smaller earthquakes that follow almost all sizable earthquakes. Their number, magnitudes, and durations are proportional to the magnitude of the principal shock.

Amplitude (wave)   The maximum departure of the value of a wave from its average value; that is the height of the wave's crest or the depth of its trough.

Arid   Term applied to an area that receives 25 to 200 mm (8 in.) of rainfall annually.

Aseismic creep   Slow movement along a fault that does not produce earthquakes.

Ash   Fragments less than 2 mm in diameter ejected from a volcano.

Asthenosphere   A layer of partially molten material underlying the *lithosphere* that has no lasting endurance to shearing stress.

Atmospheric circulation   The movement of air masses above the earth's surface.

Avalanche   A very rapid downslope movement of rock, soils, or mixtures of ice, rock, and snow.

Base shear coefficient (equivalent lateral force)   The shearing force applied at the base of a building. It is related to the percentage of the earth's gravitational *accelera-tion* that a building receives during intense ground shaking.

Benioff zone   The locus of earthquake foci paralleling the downward trend of a subducting lithospheric plate.

Bombs   Pieces of rock blown out of an erupting volcano that are larger than 64 mm in diameter.

Bracing   Construction techniques applied to a building to prevent lateral swaying and failure during an earthquake.

Brontide   A natural sound effect of an earthquake.

Causal hazard chain   An accident scenario; a predetermined sequence of hazards and potential failures, each of which has a mitigative alternative.

Core   The dense nucleus of the earth that consists of a fiery liquid mass surrounding an inner solid mass.

Coriolis effect   The apparent deflection of a moving object such as a ballistic missile or the earth's atmosphere and its oceans, to the right in the northern hemisphere and to the left in the southern hemisphere as a result of the earth's rotation.

Crust   The outer shell of the earth, some tens of kilometers in thickness.

Deep-focus earthquakes   Those that occur in the depth range of 300 to 800 km.

Desertification   Expansion of the sands of the desert; also the process whereby semiarid lands along the peripheries of deserts become agriculturally useless because of climatic change, degradation, or overuse.

Disaster   A calamitous event that causes enormous distruction of life and/or property.

Doldrums   The equatorial belt of light, variable winds lying between the two trade wind belts.

Drainage basin   The land area that is drained by a river and its tributaries.

Drift (of a building)   The maximum deflection from the vertical of the top of a building during ground shaking.

Drought   An extended period of deficit rainfall that results in the curtailment of the natural growth of vegetation and organisms in a region.

Dust storm   Dust-laden whirlwind that moves across an arid region.

Dust devil   A small whirlwind of dust and sand.

Earthquake   A sudden shaking in the earth caused by the release of accumulated stresses in the rocks of the earth's crust; generally takes place along a *fault*.

Ecosystem   The complex interrelationship between plants and animals and their environment that makes up a functioning biophysical unit.

Energy   The capacity of a system to do work; commonly measured in ergs.

Epicenter   The point on the earth's surface directly above the *focus* of an earthquake.

Eye   The roughly circular area of light winds and fair weather at the core of a tropical cyclone or hurricane.

Fault   A fracture or zone of weakness in a rock.

Fault offset   A horizontal or vertical displacement formed across a fault by ground movement during an earthquake.

Fetch   The geographical length of water over which wind blows and generates waves.

Flash flood   A sudden local flood of great volume and short duration generally resulting from heavy rainfall in the immediate vicinity.

Floodplain   That extent of relatively flat land adjacent to a river that is subject to flooding when the river overflows and that retains a deposit of river sediment after the waters recede.

Focus (hypocenter)   The point within the earth where an earthquake rupture begins.

Föhn (or foehn)   A warm, dry wind blowing down the leeward side of a mountain range.

Forecasting   The probability that an event, such as an earthquake of a specified size, will occur within a considerable time span. It is less specific than a *prediction*.

Hazard   A source of danger whose evaluation encompasses three elements: risk of personal harm, risk of property damage, and the acceptability of the level or degree of risk.

Hurricane   See *tropical cyclone*.

Hyperarid   Term applied to an area that receives less than 25 mm (1 in.) of rainfall annually.

Iceberg   Massive floating body of ice that has broken off from a glacier into the ocean.

Intermediate-focus earthquakes   Those that occur in the depth range from 70 to 300 km.

Interplate earthquakes   Earthquakes that occur along the boundaries of interacting lithospheric plates.

Intraplate earthquakes   Earthquakes that occur within the interior of lithospheric plates.

Isoseismal map   A contour map depicting regions of equal seismic intensity.

Lahar   Mudflow containing rock debris and blocks of volcanic origin mixed with water; produced by the mixing of volcanic ejecta with ice or snow covering the volcano's slope.

Landslide   The movement of a mass of earth material downslope.

Lapilli   Volcanic ejecta ranging in diameter from 2 to 60 mm.

Lava   *Magma* that has reached the earth's surface through a volcanic eruption.

Levee   A river embankment built higher than the adjacent floodplain to prevent flooding.

Liquefaction   The process whereby a mixture of soil and sand behaves like a fluid rather than a wet solid mass during earthquake shaking.

Lithosphere   The somewhat rigid shell of the earth, consisting of the *crust* and upper *mantle*, that overlies the mobile *asthenosphere*. All earthquakes occur in the lithosphere.

Magnitude (earthquake)   A measure of earthquake size based on the logarithm to the base 10 of the maximum departure of seismic waves from their average value, as registered on a seismograph and corrected for the distance to the epicenter. Three types of magnitude scales are in use: Richter or local magnitude and two scales based on *amplitudes* seen on different portions of a seismic record.

Magma   Molten rock deep within the earth's *crust*; when it reaches the surface, it is called *lava*.

Mantle   The solid intermediate layer of the earth, between the crust and the core, that has a depth ranging from tens of kilometers to 2900 km.

M-discontinuity   The boundary between the earth's crust and mantle that is marked by an increase in speed of propagation of seismic waves.

Meander   The tendency of a river to travel in a winding path.

Mitigation   The minimization action of the effects of a natural hazard; it has physical, engineering, and social aspects.

Nuée ardente   A glowing avalanche of gas and volcanic ash that flows downslope at a great velocity.

Pack ice   Pieces of floating sea ice driven together into a single mass.

Pahoehoe   The Hawaiian term for *lava* characterized by a smooth, ropy, or billowy surface.

Plates   The large movable segments into which the earth's surface is divided and which underlie the continents and the oceans.

Prediction   The specification of the time, place, magnitude, and probability of occurrence of an event such as an earthquake.

Pyroclastics   Fragmentary rocks ejected from a volcanic vent, often in great volume.

Recurrence interval   The average time between occurrences of an event of a given magnitude (e.g., between eruptions of a volcano.)

Safety threshold   The point above which a phenomenon constitutes a hazard.

Scarp   A cliff or near-vertical slope formed along a fault by ground movement during an earthquake.

Seiche   Oscillation of an enclosed or semi-enclosed body of water produced by an earthquake, an atmospheric pressure disturbance, or strong winds.

Seismic activity   The sudden movement in the earth caused by an earthquake.

Seismic gap   The portion of a plate boundary that has not been ruptured by earthquake activity for some time.

Seismic intensity   A numerical measure of ground shaking on a scale of I to XII. The scale is based on observations of damage to structures and on reports of shaking felt by inhabitants.

Seismicity   The relative frequency and distribution of earthquakes.

Seismic moment   The surface area of the fault area being displaced multiplied by the average displacement distance and the rigidity of the rocks involved.

Seismic sea wave   See *tsunami*.

Seismograph   An instrument for recording earthquake disturbances.

Semiarid   Term applied to areas that receive 200 to 500 mm (20 in.) of rainfall annually.

Shallow-focus earthquakes   Those that take place within 70 km of the earth's surface.

Shearing resistance   The horizontal frictional resistance at the base of a structure opposing the applied horizontal acceleration of a tremor.

Slip   The relative displacement of one side of a fault relative to the other after an earthquake.

Slip rate   The rate of movement between lithospheric plate boundaries.

Storm surge   The transient motion, created by strong winds, of a portion of a body of water that results in a mean water level at the shoreline that is above the normal tidal level of oscillation.

Subduction zone   Formed when a thinner oceanic plate converges with a thicker continental plate and is forced or subducted beneath the continental plate.

Summer solstice   The time in late June when the vertical rays of the sun attack at $23\frac{1}{2}°$ north latitude (the *Tropic of Cancer*).

Tephra   General term for all types of material ejected by a volcano, including *pyroclastics*.

Tornado   A violently rotating column of air.

Trade winds   An extremely consistent set of winds occupying most of the tropics, and a major component of the global pattern of atmospheric circulation. In the northern hemisphere they flow northeasterly between the 20° and 30° latitudes; in the southern hemisphere they flow southeasterly within the same latitudes.

Tropical cyclone   A storm that rotates about a center of low atmospheric pressure and has wind speed of at least 119 km/hr. Called a *hurricane* in the western Atlantic Ocean and Caribbean and a *typhoon* in the western Pacific.

Tropic of Cancer   The parallel of latitude that is at $23\frac{1}{2}°$ north of the equator and that is the northernmost latitude reached by the sun's vertical rays.

Tropic of Capricorn   The parallel of latitude that is at $23\frac{1}{2}°$ south of the equator and that is the southernmost latitude reached by the sun's vertical rays.

Tsunami   A large, long water wave produced by a sudden change in the elevation of a seabed resulting from an earthquake, a volcanic eruption, or a submarine landslide. Also called a *seismic sea wave*.

Typhoon   See *tropical cyclone*.

Volcano   A vent in the earth's crust from which *lava* is ejected.

Vortex   A swirling or spinning mass of fluid that can achieve considerable velocity.

Water bore   A tsunami amplified, in a bay, river mouth, or other funnel-shaped inlet, into a near-vertical wall of water.

Wave diffraction   The modification a water wave undergoes as a result of limitation of its lateral extent by a breakwater.

Wave frequency   The number of times that a wave crest or trough passes a fixed point within a specified interval of time. Frequency is the reciprocal of *wave period*.

Wavelength   The horizontal distance between two successive crests (peaks) or troughs of a wave.

Wave period   The interval of time between the passage of successive crests or troughs of a wave.

Wave refraction   The change in direction of a propagating wave—either seismic or water—as it encounters a region of different wave speed or velocity.

Wave shoaling   The movement of a water wave from a greater to a lesser depth of water.

Wave velocity   The wave's speed plus the direction of its approach.

Weather   The temporary and localized state of the atmosphere, measured on a time scale of hours to days.

Weir   An obstruction placed across a river to raise its level or divert its flow.

Winter solstice   The time in late December when the vertical rays of the sun attack at $23\frac{1}{2}°$ south latitude (the *Tropic of Capricorn*).

Bibliography

American Iron and Steel Institute. 1962. *The Agadir, Morocco Earthquake, February 29, 1960*. New York: American Iron and Steel Institute.

Applied Technology Council, ATC 3-06. 1978. *Tentative Provisions for the Development of Seismic Regulations for Buildings*. Redwood City, Calif.: Applied Technology Council.

Baker, V. R., R. C. Kochel, and P. C. Patton. 1988. *Flood Geomorphology*. New York: Wiley.

Bayer, K. C., L. E. Heuckroth, and R. A. Karim. 1969. "An Investigation of the Dasht-E. Bayāz, Iran Earthquake of August 31, 1968." *Bull. Seism. Soc. Am. 59*: 1793–1822.

Bick, T., C. Hohenemser, and R. Kates. 1979. "Target: Highway Risks." *Environment 21*: 7–15.

Bogard, W. 1989. *The Bhopal Tragedy*. Boulder, Colo.: Westview.

Bolt, B. A. 1993. *Earthquakes*. New York: Freeman.

Bolt, B. A., W. L. Horn, G. A. MacDonald, and R. F. Scott. 1977. *Geological Hazards*. New York: Springer-Verlag.

Brabb, E. E., and B. L. Harrod, eds. 1989. *Landslides: Extent and Economic Significance*. Rotterdam: Balkema.

Brookes, I. A. 1987. "A Medieval Catastrophic Flood in Central West Iran." In *Catastrophic Flooding*, ed. L. Mayer and D. Nash. London: Allen and Unwin. Pp. 225–46.

Bryant, E. A. 1991. *Natural Hazards*. Cambridge: Cambridge University Press.

Bullard, F. M. 1976. *Volcanoes of the Earth*. Austin: University of Texas Press.

Burton, I., R. W. Kates, and G. F. White. 1993. *The Environment as Hazard*. 2nd ed. New York: Guilford Press.

Caldwell, J. C. 1984. "Desertification." Occasional Paper No. 37, Development Studies Centre, Australian National University. 1–50 pp.

Chernousenko, V. M. 1991. *Chernobyl—Insight from the Inside*. Berlin: Springer-Verlag.

Coburn, A., and R. Spence. 1992. *Earthquake Protection*. Chichester, England: Wiley.

Command of the Viceroy. 1748. *A True and Particular Relation of the Dreadful Earthquake which happen'd at Lima, the Capital of Peru, and the Neighboring Port of Callao, on the 28th of October, 1746*. London: Osborne.

Crozier, M. J. 1984. "*Field Assessment of Slope Instability*." In *Slope Instability*, ed. D. Brunsden and D. Prior. New York: Wiley. Pp. 103–42.

Crozier, M. J. 1986. *Landslides: Causes, Consequences and Environment*. London: Croom Helm.

Czaya, E. 1983. *Rivers of the World*. New York: Van Nostrand Reinhold.

Darkoh, M. B. K. 1980. *Man and Desertification in Tropical Africa*. Inaugural Lecture Series, No. 26. Dar Es Salaam. 1–51 pp.

Derr, J. S. 1973. "Earthquake Lights: A Review of Observations and Present Theories." *Bull. Seism. Soc. Amer. 63*: 2177–87.

Dregne, H. E. 1983. *Desertification of Arid Lands*. Chichester, England: Harwood.

Eagleman, J. R. 1983. *Severe and Unusual Weather*. New York: Van Nostrand Reinhold.

El-Sabh, M. I., and T. S. Murty, eds. 1988. *Natural and Man-Made Hazards*. Dordrecht: Reidel.

Freeman, J. R. 1932. *Earthquake Damage and Earthquake Insurance*. New York: McGraw-Hill.

Gardner, J. 1970. "Rockfall: A Geomorphic Process in High Mountain Terrain." *Albertan Geogr. 6*: 15–20.

Grainger, A. 1990. *The Threatening Desert*. London: Earthscan.

Gross, M. G. 1987. *Oceanography: A View of the Earth*, 4th ed. Englewood Cliffs, N.J.: Prentice Hall.

Grove, A. T. 1978. *Africa*. Oxford: Oxford University Press.

Gutenberg, B., and C. F. Richter. 1954. *Seismicity of the Earth and Associated Phenomena*. Princeton, N.J.: Princeton University Press.

Healy, J. H., V. G. Kossobokov, and J. W. Dewey. 1992. *A Test to Evaluate the Earthquake Prediction Algorithm, M8*. U.S. Geological Survey Open File Report 92–401.

Healy, J. H., W. W. Rubey, D. T. Griggs, and C. B. Raleigh. 1968. "The Denver Earthquakes." *Science 161*: 1301–10.

Hoyt, W. G., and W. B. Langbein. 1955. *Floods*. Princeton, N.J.: Princeton University Press.

Keeney, R. 1980. *Siting Energy Facilities*. New York: Academic Press.

Kendrick, T. D. 1956. *The Lisbon Earthquake*. London: Metheun.

Kochel, R. C. 1988. "Geomorphic Impact of Large Floods: Review and New Perspectives on Magnitude and Frequency." In *Flood Geomorphology*, ed. V. R. Baker, R. C. Kochel, and P. C. Patton. New York: Wiley. Pp. 169–87.

Kovach, R. L. 1988. "Earthquake Hazard in Jordan." *Natural Hazards 1*: 245–54.

Kovach, R. L., and J. H. Healy. 1990. *Jordan Seismic System*. U.S. Geological Survey Open File Report 90–667. 138 pp.

Lacroix, A. 1904. *La Montagne Pelée et ses Eruptions*. Paris: Masson.

Lagorio, H. 1991. *Earthquakes: An Architect's Guide to Non-Structural Seismic Hazards*. New York: Wiley.

Lutgens, F. K., and E. S. Tarbuck. 1992. *The Atmosphere*. Englewood Cliffs, N.J.: Prentice Hall.

Mainguet, M. 1991. *Desertification: Natural Background and Human Mismanagement*. Berlin: Springer-Verlag.

Medvedev, S. V. 1968. "The International Scale of Seismic Intensity." In *Seismic Zones of the USSR*. Moscow: Nauka.

Meister, L. J., R. O. Burford, G. A. Thompson, and R. L. Kovach. 1968. "Surface Strain Changes and Strain Energy Release in the Dixie Valley-Fairview Peak Area, Nevada." *J. Geophys. Res. 73*: 5981–94.

Officers of the Geological Survey of India and S. C. Roy. 1939. "The Bihar-Nepal Earthquake of 1934." *Mem. Geol. Survey India 73*: 1–391.

Oldham, R. D. 1899. "Report on the Great Earthquake of 12th June 1897." *Mem. Geol. Survey India 29*: 1–379.

O'Riordan, T. 1986. "Coping with Environmental Hazards." In *Geography, Resources and Environment*, ed. R. W. Kates and I. Burton. Chicago: University of Chicago Press. Pp. 272–309.

Perret, F. A. 1924. *The Vesuvius Eruption of 1906: Study of a Volcanic Cycle*. Carnegie Institution of Washington, Publ. No. 339.

Reiter, L. 1990. *Earthquake Hazard Analysis*. New York: Columbia University Press.

Richards, K. 1982. *Rivers, Form and Process in Alluvial Channels*. London: Methuen.

Richter, C. F. 1958. *Elementary Seismology*. New York: Freeman.

Scandone, R. 1983. "Problems Related with the Evaluation of Volcanic Risk." In *Forecasting Volcanic Events*. Amsterdam: Elsevier. Pp. 57–67.

Starr, C. 1969. "Social Benefit versus Technological Risk: What Is Our Society Willing to Pay for Safety?" *Science 165*: 1232–36

Stierman, D. J. 1980. "Earthquake Sounds and Animal Cues." *Bull. Seism. Soc. Am. 70*: 639–43.

Symons, G. J. 1888. *The Eruption of Krakatoa and Subsequent Phenomena*. London: Royal Society of London.

Thomson, W. 1880. *The Land and the Book*. New York: Harper.

Tilling, R. I. 1989. "Volcanic Hazards and Their Mitigation: Progress and Problems." *Reviews of Geophysics 27*: 237–69.

Varnes, D. J. 1958. "Landslide Types and Processes." In *Landslides and Engineering Practice*, ed. E. B. Eckel. Washington, D.C.: Highway Research Board Special Report 29. Pp. 20–47.

Wemelsfelder, P. J. 1961. "On the Use of Frequency Curves of Stormfloods." *Proc. Seventh Conf. Coastal Engineering* (Engineering Foundation Council on Wave Research, A.S.C.E.): 617–32.

Whipple, A. B. C. 1982. *Storm*. Alexandria, Va.: Time-Life.

White, G. F., ed. 1974. *Natural Hazards: Local, National, Global*. New York: Oxford University Press.

Wiegel, R. L. 1964. *Oceanographical Engineering*. Englewood Cliffs, N.J.: Prentice Hall.

Wiegel, R. L. 1970. *Earthquake Engineering*. Englewood Cliffs, N.J.: Prentice Hall.

Wijkman, A., and L. Timerlake. 1984. *Natural Disasters: Acts of God or Acts of Man?* London: Earthscan.

Wilson, R. 1979. "Analyzing the Daily Risks of Life." *Technology Review 81*: 40–46.

Verbeek, R. D. M. 1886. *Krakatau*. Batavia: Imprimerie de l'état.

Yanev, P. 1974. *Peace of Mind in Earthquake Country: How to Save Your Home and Life*. San Francisco: Chronicle.

Zonghu, Z., Z. Zhiyi, and W. Yunsheng. 1991. *Loess Deposit in China*. Beijing: Geological Publishing House.

Answers
to Review Questions

CHAPTER 1

1. What is a hazard?

Answer: A hazard is a source of danger whose evaluation encompasses three elements: risk of human harm, risk of property damage, and the acceptability of the level or degree of risk.

---

2. List the considerations necessary for evaluating a potential natural hazard.

Answer:
   (a) Perception of a phenomenon
   (b) Identification of the phenomenon as a hazard
   (c) Risk evaluation
   (d) Assessment and control

---

3. Why is the setting of a safety threshold for a hazard important? What is the chief danger in setting a catastrophic threshold of risk?

Answer: Setting a safety threshold is equivalent to specifying the level of risk that is socially acceptable. If the catastrophic threshold of risk is set unduly high, the ability to detect lower-level failures or risks is compromised.

4. Which two sudden-onset natural hazards produce, on an annualized basis the greatest fatality rates?

Answer: On an annualized basis, earthquakes and tropical cyclones cause the most deaths.

CHAPTER 2

1. What were lessons learned from the eruption of Mount Pinatubo in the Philippine Islands?

Answer: Increased unrest observed at a volcano long thought to be dormant must be heeded; uncertain forecasts are valuable for the assessment of preparedness; time-sequential upgraded levels of monitoring of volcanoes are needed; and the past eruption history of a volcano needs to be carefully scrutinized.

2. What is the difference between magma and lava?

Answer: Magma is molten rock underground while lava is molten rock that has flowed out of a volcano onto the earth's surface.

3. Distinguish among active, dormant, and extinct volcanoes.

Answer: Active or live volcanoes erupt regularly or continually. Dormant volcanoes erupt less frequently. Extinct, or dead, volcanoes are presumed to be incapable of erupting again. The distinction between *dormant* and *extinct* is difficult to make with assurance.

4. List and describe the major types of hazardous volcanic events.

Answer: *Pyroclastic (tephra) flow*: Volcanic eruption consisting of fragments of rock and partially or wholly solidified lava that range in size from ash (< 2 mm) to lapilli (2–64 mm) to bombs (> 64 mm). *Pyroclastic flow*: Volcanic expulsion of incandescently heated mixture of gas and rock that may produce a nuée ardente, which is a scalding avalanche of gas and solid material, or a debris avalanche. *Lahar*: Mudflow consisting of rock debris and blocks predominantly volcanic in origin mixed with water. *Lava flow*: Slow-moving mass of molten rock at the surface. *Volcanic gases*: Toxic gases, such as CO, CO_2, H_2S, and SO_2, sometimes emitted by a volcano.

---

5. What types of lava are found on the island of Hawaii?

Answer: The two types of Hawaiian lava flows are smooth, billowly pahoehoe and jagged and clinkery aa.

---

6. What is the most dangerous type of volcanic event in terms of human fatalities?

Answer: Pyroclastic flows are the *direct* cause of more fatalities than any other volcanic event, though inadequate responses to *all* kinds of volcanic disasters actually result in much higher death tolls, through disease and starvation, than any type of volcanic action in itself.

---

7. Why do people choose to live near volcanoes?

Answer: Volcanic ash produces fertile farmland, and volcanic rocks are excellent building materials; some volcanic areas provide geothermal energy; the higher elevations of volcanic areas means cooler weather in the tropics and subtropics; population pressures; scenic beauty.

---

8. Name the four most important considerations for volcanic hazard assessment.

Answer: Volcano status, history, recurrence interval, and the concept of volcano hazard zonation.

---

9. Describe the concept of volcanic hazard zonation.

Answer: Volcanic hazard zonation classifies areas according to the degree of risk they pose for volcanic eruption. The key considerations in doing this type of zoning are: number of people exposed, their locations, and their economic activities; the location of infrastructure and emergency services and their susceptibility to hazard; and potential crop loss in agricultural areas.

CHAPTER 3

1. Where do 80% of the world's earthquakes occur? What smaller region of the world contains 15% of earthquake activity in terms of energy release?

Answer: 80% of the world's earthquakes take place in the circum-Pacific belt (the "Ring of Fire"). The arc circling the Aegean Sea that includes the countries of Greece, Albania, Romania, and Italy accounts for about 15% of global seismic energy release.

---

2. Distinguish between an earthquake's epicenter and its hypocenter (focus).

Answer: The hypocenter or focus is that point within the earth where the earthquake rupture starts, while the epicenter is that point on the earth's surface that lies directly above the focus.

---

3. Distinguish between lithosphere and asthenosphere. In which of these layers do earthquakes occur? Aseismic creep?

Answer: The lithosphere is the strong, rocky, somewhat rigid shell of the earth consisting of the crust and a portion of the upper mantle. It is the material of which the plates are made. Beneath the continents, lithospheric thickness ranges from 100 to 200 km and beneath the oceans it is about 70 km. The asthenosphere is a layer of hot, partly molten material underlying the lithosphere that has no lasting endurance to shearing stress. All earthquakes occur in the lithosphere. Aseismic creep, slow movement that does *not* produce earthquakes, occurs in the asthenosphere.

---

4. Why do shallow-focus earthquakes occur along oceanic ridges (accretionary plate boundaries)?

Answer: Shallow-focus earthquakes take place here because of divergent or differential movement along these kinds of boundaries. As midoceanic ridges are pulled apart or slip along zones of weakness, material from below is injected into the gap between boundaries to form new oceanic floor.

---

5. Where are intermediate- and deep-focus earthquakes commonly located?

Answer: Intermediate- and deep-focus earthquakes commonly occur in the Benioff zone, that area along the boundaries or within the interior of subducting plates. The locus of earthquake foci parallels the downward trend of the subducting plate.

---

6. What is the difference between a scarp and a horizontal fault offset?

Answer: Both are caused by ground displacement during an earthquake. A scarp is a cliff or near *vertical* slope formed by *vertical* movement along a fault, while a horizontal offset is formed by *horizontal displacement* along a fault.

7. What is a brontide?

Answer: A brontide is a *natural* earthquake noise that usually sounds like low but loud rumbling.

---

8. What is an isoseismal map?

Answer: An isoseismal map is a contour map depicting regions of equal seismic intensity or shaking after assessments are made at various localities.

---

9. What are the three major problems in assessing earthquake intensity?

Answer: Local geologic effects often receive undue emphasis; there is usually a bias towards inhabited areas; the subjective descriptions used in assessment are frequently unreliable.

---

10. Your friend tells you, "That earthquake got me out of a deep sleep. Some plaster fell on my head and my dresser moved across the room." From his statement you know that the Modified Mercalli Intensity was at least what value?

Answer: His description is a good indication that this earthquake had at least a Mercalli Intensity of VI.

---

11. We want to estimate the magnitude of an earthquake required to completely circle the globe. We make some simplifying assumptions concerning storage of the energy released in an earthquake. Assume that the energy is stored in a box surrounding the fault.

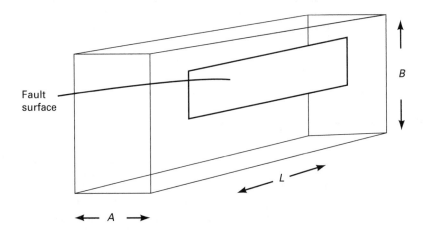

The total energy released is approximately $E = KABL$, where K represents the energy stored per unit volume. Assume that large earthquakes differ in energy release only through the fault length L (that is, assume, that K, A, and B are constant for large earthquakes). Knowing that (1) a magnitude 8.5 earthquake corresponds to $L = 400$ km ($\log 400 = 2.60$), (2) $\log E = 11.8 + 1.5$M, (3) the circumference of the earth is 40,000 km ($\log 40{,}000 = 4.60$), and (4) $\log E = \log L + \log(KAB)$, what magnitude earthquake would be required to produce a fault with L equal to the circumference of the earth?

Answer: In this problem we know that $\log E = \log L + \log(KAB)$. Since we assume that KAB is a constant, we first determine its value from a (400 km) magnitude 8.5 earthquake of length L and energy E:

$$\log E = 11.8 + 1.5\, M$$

so that $11.8 + 1.5M - \log L = \log(KAB)$

with our values

$$11.8 + 1.5\,(8.5) - \log 400 = \log(KAB)$$

giving

$$\log(KAB) = 21.95$$

We now know that

$$\log E = 40{,}000 + 21.95 = 26.55$$

and

$$\log E = 26.55 = 11.89 + 1.5\, M$$

which gives

$$M = 9.8$$

---

12. The magnitude of the 1971 San Fernando earthquake was 6.4, and the 1906 San Francisco earthquake reached an estimated magnitude of 8.3. By approximately what factor was the ground motion at San Francisco greater than that at San Fernando?

Answer: Ground motion increases by a factor of 10 for each unit increase in magnitude. Therefore the 1906 San Francisco earthquake produced ground amplitudes approximately 100 times greater than those produced by the 1971 San Fernando earthquake.

---

13. Describe three theories of earthquake prediction.

Answer: One theory uses seismic gaps—portions of plate boundaries that have not been ruptured by earthquake activity for some time—to pinpoint likely sites for future major earthquakes. The second theory uses *premonitory phenomena*—observable physical precursors to large earthquakes, including ground tilt, crustal deformation, and variations in the electrical resistance of rocks—to predict the location and magnitude of severe shocks. The third theory, based on the hypothesis that increases in small and medium-sized shocks in a broad region will precede a great earthquake, uses a computer algorithm to identify *times of increased probability* (TIPS) of earthquake occurrence in the circum-Pacific belt.

CHAPTER 4

1. What is the drift of a building?

Answer: The drift of a building is the maximum deflection from the vertical of the top of a building produced by oscillatory earthquake motion.

---

2. What are the key principles for earthquake-resistant construction?

Answer: The key principles for earthquake-resistant construction are symmetry and regularity. Also important are bracing, attachment of load-bearing walls to a substantial foundation, and building on suitable ground.

---

3. In the Romanian earthquake of March 4, 1977, most collapsed buildings in Bucharest were 10- to 12-story concrete frame buildings on block corners. Why do you think corner buildings are more likely to collapse than midblock buildings? (Do not assume that corner buildings are any more asymmetric or irregular than midblock buildings.)

Answer: Corner buildings are more likely to collapse because they are not supported on two adjacent sides. Therefore, corner buildings can fall (be pushed) in two different directions—for example, north-south and east-west. Midblock buildings are supported on opposite sides by their neighboring buildings, so they can only fall one way—for example, east-west. It is possible for a building to fall in only two directions in this example because the *rocking* is in one plane—either north-south or east-west, but not both.

---

4. What are the critical elements of a building code?

Answer: The critical elements of a building code are seismic risk zoning, importance factors, and drift control.

5. Why did so many buildings collapse in the 1968 Dasht-e Bayāz earthquake?

Answer: The major reasons were that these buildings were made of weak materials, had no reinforcement, and lacked foundations.

6. What is the most important objective of microregionalization?

Answer: The most important objective of microregionalization or microzoning is the quantification of local geological site characteristics so that estimates of future seismic intensities and ground accelerations can be made.

7. What is meant by base shear coefficient?

Answer: The base shear coefficient is a parameter relating to the level of horizontal acceleration applied to the base of a structure.

8. A building 20 stories high is subjected to horizontal acceleration at its base of 0.07 g or 7% g. Making use of the relation for the Uniform Building Code $V_0 = ZIKCSW$, what is the minimum value of K that should be specified so that the building can tolerate stresses produced by an acceleration of 7% g?

Answer: The necessary equation is

$$KC = \frac{S_A}{g}$$

You are given that $S_A/g = 0.07$, so now you must find the value of C. C is related to the period of oscillation of the building, T, by

In our case,

$$T = (0.1)(\text{number of stories}) = (0.1)(20) = 2$$

Now the problem is done except for punching calculator keys.

$$K = \frac{0.07}{C}$$

9. What is a safe-shutdown earthquake?

Answer: A safe-shutdown earthquake is the maximum-credible-magnitude earth-quake that a nuclear power plant must withstand without any loss in structural in-tegrity.

10. What are the key earthquake-bracing techniques for a building?

Answer: The main earthquake-bracing techniques for buildings are frame-action, shear-wall, and diagonal or "X" bracing.

11. What lessons were not learned after the 1906 San Francisco earthquake?

Answer: That a reliable water distribution system is critical and that fire-resistant construction is necessary in an earthquake-prone region.

CHAPTER 5

1. What is the difference between a mudflow and a lahar?

Answer: A mudflow is a moving mass of soil whose water content (from rain or snow) is the major factor controlling its downslope movement. A lahar is kind of a mudflow, but instead of ordinary rocks, soil, and fill, it is composed of *volcanic ash* and *gases*, mixed with melted ice or snow from the volcano's slope.

2. Why is it difficult to determine damage done by landslides per se?

Answer: The difficulty in assessing the damaging effects of landslides per se is that it is often impossible to separate these from the effects of the event that triggered the landslide: usually an earthquake, a volcanic eruption, very heavy rains, or a thunder-storm.

3. Why does landslide damage commonly occur in hillside housing developments?

Answer: Landslide damage is common in hillside housing developments because the cut-and-fill procedure used to grade the land was improper. Either the cut was too steep or the fill was not sufficiently compacted.

---

4. What are the key factors for initiating downslope movement of a mass of rock and soil?

Answer: The two key factors are slope instability and a reduction in shearing resistance of the soil or rock material.

---

5. What criterion would you use to assess whether future landslides in an area have the potential for becoming sudden-onset disasters?

Answer: The most useful criterion would be velocity, or rate of downslope movement, of past landslides in the area. The more rapid the speed of previous landslides' downslope movement, the more likely a sudden-onset disaster is in the future.

---

6. Why do landslides commonly occur during the rainy season?

Answer: Landslides tend to occur during the rainy season because water reduces the soil and rock's frictional resistance to downward sliding.

---

7. What are the main ways you can mitigate the hazards of landslides and other downslope earth movements?

Answer: Avoid areas that have been geologically evaluated as regions of potential slope instability; heed official warnings to evacuate when necessary; press for sensible land-use restrictions and engineering countermeasures.

---

8. What is the chief obstacle to developing a comprehensive rapid warning system for landslides?

Answer: The chief obstacle to establishing a comprehensive rapid warning system for landslides is that there are so many areas susceptible to landslides that it is not feasible to monitor them all with geotechnical instruments.

CHAPTER 6

1. What is the meaning of desertification?

Answer: Desertification is a slow-developing environmental crisis, the process whereby semiarid lands become desertlike because of climatic change and land degradation.

---

2. Distinguish between a hyperarid region and a semiarid region.

Answer: A hyperarid area typically receives less than 25 mm of rain per year; no crops can grow in these areas, which are true deserts. Semiarid regions receive annual rainfall amounts ranging from 200 to 500 mm; some crop growing is possible in these areas.

---

3. What continent possesses the largest percentage of the dry regions of the world?

Answer: Africa has the highest percentage of dry regions, but Asia is a fairly close second.

---

4. What are the primary causes of desertification?

Answer: Land abuse exacerbated by drought, is the primary cause of desertification.

---

5. Name the major types of land abuse.

Answer: Overcultivation, overgrazing, deforestation, and bad irrigation practices are the major forms of land abuse. All lead to the deterioration of the soil and destruction of its biological capability.

---

6. What is a drought?

Answer: A drought is an extended period of rainfall deficit that leads to the curtailment of the natural rate of growth of vegetation and organisms in a region. The length of this period varies from region to region, according to the long-term rainfall *norm* for the region.

---

7. Identify the principal measures that can be taken to halt and possibly reverse desertification.

Answer: The three principal measures for halting the process of desertification are soil conservation and rejuvenation, the substitution of other fuels for wood in order to stop deforestation and replanting trees and vegetation where possible, and the discouragement of cattle herding.

8. How are off-road vehicles contributing to desertification?

Answer: Indiscriminately driven in semiarid lands, off-road vehicles produce gullies or gouges that expose the underlying bedrock to more rapid erosion by wind and water.

CHAPTER 7

1. What are the three main factors controlling the global pattern of atmospheric circulation?

Answer: The three main factors controlling the global pattern of atmospheric circulation are: the tilt of the earth's axis of rotation from the plane of its elliptical orbit around the sun, which produces differential heating; the rotation of the earth which prodces the very important Coriolis effect; and the existence of continental land masses.

2. How does a tropical cyclone originate?

Answer: A tropical cyclone develops most often in late summer when the temperature of oceans near the equator is sufficiently high to produce heat and moisture in the air, which is lifted to higher elevations. This lifting of heated air causes a tropical low-pressure area, which pulls in colder, drier air from the surrounding atmosphere. Since they are subject to the Coriolis effect, these inward flowing or centripetal winds spin—counterclockwise in the northern hemisphere and clockwise in the southern hemisphere.

3. Why does a tropical cyclone weaken when it moves outside the zone of tropical seas?

Answer: It weakens either because it hits land, where there is dry air, or because as it travels into cooler latitudes over the oceans, the cooler sea surface temperatures rob it of warm, moist air.

4. Why do tropical storms that originate within 5° of the equator not acquire as much of a rotary motion as storms or cyclones originating at somewhat higher latitudes?

Answer: Tropical storms that originate within 5° of the equator do not acquire much of a rotary motion because the Coriolis effect is weak or absent around the equator.

---

5. A tornado has higher wind speeds than a hurricane, yet a hurricane produces more total damage. Why?

Answer: A tropical cyclone produces more total damage than a tornado because its lateral dimension is much larger and it lasts much longer, and also because its torrential rains often cause storm surges and inland flooding.

---

6. Why does a tornado form?

Answer: Tornadoes form in conjunction with severe thunderstorms, usually when cold, dry air from the Arctic collides with warm, humid tropical air from the Gulf of Mexico.

---

7. When you receive a tornado warning, should you board up your windows? Explain.

Answer: No. Recall that within the center of a tornado the air (barometric) pressure drops significantly. If your house is closed up, the air pressure within will exceed that of the tornado as it passes and your house will literally explode.

---

8. What are dust storms and why do they occur?

Answer: Dust storms are wind storms carrying suspended soil and sand material without any accompanying precipitation. They are caused by a conjunction of low rainfall and land degradation from poor land use and mismanagement.

---

9. What is a dust devil?

Answer: Dust devils, commonly found in arid and semiarid regions are whirling vortexes of dust resembling tornadoes but much smaller in size. Typical dimensions are a few meters in diameter and 100 m in height. Unlike tornadoes, dust devils form from the ground up, because surface heating is critical to their formation.

---

10. Do tropical cyclones have any beneficial effects?

Answer: Tropical cyclones provide essential rainfall over many areas that they cross. Japan, for instance, depends on tropical cyclones for 25% of its rainfall.

CHAPTER 8

1. Why does a change in the speed of a water wave approaching a shoreline produce an increase in hazard?

Answer: As a wave approaches a shoreline, its speed decreases, compressing its wavelength. Because energy always remains constant in a column of water, the compression of wavelength results in an increase in wave height. Water is not a rigid column of material, so these higher waves curl over and crash into the shoreline.

2. What causes waves in the open ocean?

Answer: Waves generate in the open ocean because of the transference of energy from winds to the body of water. Since winds vary in intensity, there are pressure differences over the water. These pressure differences produces local variations in the level of water—that is, waves.

3. Why is the shape of a coastline an important factor for assessing potential damage from waves?

Answer: A coastline's shape influences the "angle of attack" of a water wave. Where there is a headland or promontory, wave refraction focuses on this land extension and away from the broader adjacent coastline, producing more forceful waves. Narrow channels also focus waves. In both cases, the angle of attack is greatly accentuated when there are storm waves, so damage is likely to be much higher than at coastlines whose contours are more even.

4. What is fetch?

Answer: Fetch is the geographical length of water over which wind blows to produce waves. The longer the fetch, the bigger the resultant water wave.

5. What is the typical wave speed of a tsunami?

Answer: A tsunami's speed varies according to the depth of the water. In the open seas of the Pacific (where most tsunami develop) the speed of a typical tsunami is 700 km/hr. As a tsunami approaches a shoreline, its speed (like that of all waves) decreases and its wave height increases.

---

6. Why is a tsunami not felt by ships at sea?

Answer: Tsunami are not felt by ships at sea because their very long offshore wavelengths make their height insignificant to the dimension of the ship itself.

---

7. A tsunami was generated by the violent explosion of Krakatoa volcano. Why?

Answer: When Krakatoa erupted, it produced a sudden vertical offset of the sea floor that resulted in a local depression in the sea level. Water rushed in to fill the depression, then moved out at right angles, the oscillations creating long-period water waves, or tsunami, that raced across the seas for thousands of miles.

---

8. What is a seiche?

Answer: A seiche is oscillation produced in a confined body of water by an earthquake, an atmospheric pressure disturbance, or strong winds. The body of water may be closed (a lake) or semiclosed (a bay).

---

9. Explain how the tsunami warning system established in Honolulu operates.

Answer: When a circum-Pacific earthquake of sufficient size is detected and located at an observatory station in the United States, Japan, Taiwan, Chile, New Zealand, or elsewhere, the data are relayed to the Seismic Sea Wave Warning Center in Honolulu, which assesses them and issues a tsunami warning, when warranted, to countries all around the Pacific.

---

10. Suppose that between 1900 and 1980 two storm surges with heights of 1.2 m were recorded at the Florida coastline. Next suppose that five storm surges with a height of 0.6 m were observed there within this same period. Calculate how many times a 1-m surge should have occurred within this time frame.

Answer: One of the observations of natural hazard sequences is that when the number of times an event occurs is plotted logarithmically against the size, height, magnitude (etc.) of the event the result is approximately linear. We will utilize the *number* of occurrences within the time interval (1900 to 1980) when specific storm surge heights were recorded to find the answer.

Height (m)	No. of Occurrences (N)	$\log_{10} N$
0.6	5	0.7
1.0	?	
1.2	2	0.3

By interpolation of the $\log_{10} N$ column we see that a 1 m surge height would be expected to have a $\log_{10} N$ value of 0.43. This translates to an N value of 2.7 times.

11. What are the chief factors in the formation of storm surges?

Answer: The chief factors in the formation of storm surges are wind setup (the speed, duration, and direction of wind) and decreases in atmospheric (barometric) pressure (which allow the mean sea level to rise). The height of a storm surge is controlled by the slope of the sea bottom.

12. Why should attempts to forecast future storm surge heights based on recurrence data from the past be looked at critically?

Answer: Since these data have usually been collected for only a short time (less than a century) and may be incomplete for even this brief period, extrapolations from them should be looked at with a careful eye. Even one missed event in the historical record can badly skew results, and the historical record is very, very brief compared to geologic time.

CHAPTER 9

1. What do the accounts of an ancient Great Deluge found in the Bible and in the Epic of Gilgamesh suggest in their similarities?

Answer: It seems likely that there *was* an unparalled flood in Mesopotamia several millenia ago that was remembered and mythologized. Both accounts suggest that there was great destruction of human life and that only a few forewarned people survived. This leads to the speculation that whatever caused the flood may have been, at least partly, human-engineered.

2. What are the two primary causes of river flooding?

Answer: The primary causes are atmospheric effects (weather fronts stationary over a river basin, causing unceasing rains) and relatively impermeable ground with little vegetation.

3. If a flood peak has a 20-year return time, what is the probability that this height will be exceeded in one year?

Answer: The reciprocal of the mean recurrence time for a particular flood height gives the probability of achieving that flood height. A flood peak height with a 20-year return time has a probability of 0.05, or 5%, of being exceeded in any given year.

4. If a particular flood peak height has a 50% probability of being exceeded in one year, what is the average or mean return time of this flood height?

Answer: If the particular flood height has a probability of 50% of being exceeded in any one year, its average return time is 2 years.

5. Why is the Hwang-Ho River constantly raising its level?

Answer: The Hwang-Ho River is constantly raising its level because the continual deposition of a fine-grained sand called *loess* is increasing the height of the river bed.

6. Describe the genesis and development of the Great Flood of 1993.

Answer: Months of extraordinary rainfall over the Mississippi River basin swelled the river and its tributaries so that they were in spate simultaneously for the first time in recorded history. The waters overwhelmed the engineering system constructed to prevent floods. By putting so many obstructions (dams, locks, and so forth) in the river's way, and by narrowing it so rigidly with levees, the system itself became part of the problem, raising the waters and channeling them with increasing force downstream.

7. Why is flash flooding often so severe in urban areas?

Answer: Flash flooding can be severe in urban areas because much of the ground is paved over, which fixes the drainage channels.

8. Name two beneficial aspects of river flooding.

Answer: River flooding provides needed irrigation for agricultural land in dry climates. In any climate its deposits of sediments produce very fertile soil.

9. List the five major steps in an effective program for the prevention and mitigation of river flooding.

Answer:

1. Recognition of the seriousness of the problem
2. Accurate risk assessment
3. Reliable forecasting, warning, and emergency response systems
4. Sound flood-control program
5. Land-use controls

CHAPTER 10

1. What is a causal hazard chain and what are its shortcomings?

Answer: A causal hazard chain is an accident scenario—a sequence of hazards and potential failures or disasters, each of which has a mitigative alternative. Its basic shortcoming is that this strategy limits mitigations to a *predetermined* set of potential failures.

2. A LNG tanker is anchored off the California coastline. An earthquake in Japan generates a tsunami that subsequently slams into the anchored tanker, producing a spill. Assume the annual probability of this disaster is 0.01% and that there is a 99% probability of no immediate ignition, so a vapor cloud forms. There is a 90% probability of an onshore wind, and you are one of 4000 people on the beach. There is a 50% probability that the first person who lights a cigarette will ignite the cloud and 2000 fatalities will result. What is the annual societal risk?

Answer: Annual societal risk is the annual probability of an accident scenario occurring times the expected number of fatalities.

$$\text{Societal risk} = (0.0001)(0.99)(0.90)(0.50)(2000)$$

$$= 0.0891$$

The annual societal risk is 8.9%.

3. What is your individual risk in the above accident scenario?

Answer: Individual risk is obtained by dividing the annual societal risk by the *total* number of people exposed. In this case, the individual risk is

$$\frac{0.0891}{4000}$$

or

$$2.2 \times 10^{-5}$$

---

4. Is the risk acceptable? Explain.

Answer: The conservative estimate for individual risk acceptability is 1 part in 10 million (1×10^{-7}). This value is much larger, so the risk is *not* acceptable.

---

5. You are one of 20 water skiers being towed by your own speedboat with its own driver on a reclaimed lake in the Sierras that is sided at the upper end by a very unstable pile of mine tailings that could easily be triggered by an earthquake, which would fill the lake and bury you. You and your friends water ski daily all year, rain or shine, and are the expected fatalities. Eighty perennial campers, skinny dippers, and other associated types on the shoreline are also exposed to the risk. If your individual annual risk is 0.01, what is the annual probability that the fatal landslide will be triggered?

Answer: In this problem there are 20 waterskiers, each with his or her own boat and driver. That means there are 40 people on the water plus 80 individuals on the shoreline. The total number of people exposed is therefore 120. The annual probability of the fatal accident is

$$\text{Individual risk} = \text{probability of accident} \times \frac{\text{fatalities}}{\text{people exposed}}$$

$$0.01 = P(A) \times \frac{40}{120}$$

The probability of the accident is 0.03 or 3%.

---

6. Suppose that a large earthquake has a return time of 24 hours and has a 40% probability of collapsing the building that you are in. You have a 30% probability of surviving the collapse. What is the overall risk of your not surviving the next 3 hours?

Answer: The overall risk is the product of the seismic risk times the probability of the various consequences.

$$\text{Risk} = [(1 - \exp(-\Delta T/T)](0.4)(1 - 0.3)$$

$$\Delta T = 3 \text{ hours}$$

$$T = 24 \text{ hours}$$

$$\text{Risk} = [\, 1 - \exp(-3/24)](0.4)(0.7)$$

$$\text{Risk} = 0.032 \quad \text{or} \quad 3.2\%$$

7. Is the annual probability of being a fatality in an earthquake disaster greater in California or Iran? Explain.

Answer: The annual probability of being a fatality from an earthquake in Iran is 1 in 23,000. In California, the probability is 1 in 2,000,000. The type of building construction must be a factor.

---

8. What is the difference between group risk and individual risk?

Answer: Individual risk is the societal risk divided by the total number of people exposed to the risk. Group risk is the probability that a member of a specified group of individuals will be a fatality.

---

9. You are a student residing in a dormitory who is paranoid about earthquakes. You know that an $M = 7$ earthquake is anticipated within a year on a nearby fault. You are told that for your area the earthquake magnitude-frequency relation is

$$\log_{10} N = 4.23 - 0.815M$$

Under the following list of provisos, what is your annual risk of dying in your dormitory?

- You have opted for year-round residence.
- You spend half of your day in your dormitory.
- The probability of your dormitory collapsing from an $M = 7$ quake is 50%.
- There is a 30% chance of surviving the collapse.
- There is a 1% chance of a subsequent nonfatal fire.

Answer: First, we need to know how often an $M = 7$ earthquake is expected to occur in our region. We calculate the number of $M = 7$ earthquakes per year, N, and then invert this to get a recurrence time, T.

$$\log_{10} N = 4.23 - (0.815)(7)$$

$$\log_{10} N = 4.23 - 5.705$$

$$\log_{10} N = -1.475$$

$$N = 10^{(-1.475)}$$

$$N = 0.0335 \text{ (earthquakes per year)}$$

$$T = 1/N = 29.8 \text{ years}$$

Next we calculate the probability of one or more $M = 7$ earthquakes occurring during the next year. Actually, because the chance of two or more $M = 7$ earthquakes occurring in the same year is so small, we could save time here and use 0.0335.

$$P = 1 - \exp \left(\frac{-1}{29.8}\right) = 0.0329$$

The probability of dying in a collapse is simply the product of the probabilities.

$$P = \begin{pmatrix} \text{probability} \\ \text{of an earthquake} \\ \text{occurring} \end{pmatrix} \begin{pmatrix} \text{probability of your} \\ \text{being inside} \\ \text{dormitory at the time} \end{pmatrix} \begin{pmatrix} \text{probability} \\ \text{of dormitory} \\ \text{collapsing} \end{pmatrix} \begin{pmatrix} \text{probability of} \\ \text{your dying in} \\ \text{the collapse} \end{pmatrix} \begin{pmatrix} \text{probability} \\ \text{of a} \\ \text{fatal fire} \end{pmatrix}$$

$$P = (0.0329)(0.5)(0.5)(1 - 0.30)(1 - 0.01)$$

$$P = 0.0057$$

$$P = 0.57\%$$

---

10. In the landslide-riggering problem discussed in the text, the San Andreas fault was determined to have a seismic risk of 6% of within 30 years. If this risk is actually 100%, what is the overall probability that a landslide would be triggered?

Answer: The individual slide probabilities during our 30-year design life are

$$P_{SA} = 1.0 \times 0.16 = 0.16$$

$$P_{MA} = 0.007 \times 0.20 = 0.001$$

$$P_{HR} = 0.08 \times 0.03 = 0.002$$

So the overall probability of a slide being triggered is now

$$P = 1 - [(1 - 0.16)(1 - 0.001)(1 - 0.002)] = 0.16 \quad \text{or} \quad 16\%$$

---

11. Under what conditions could the above overall probability be ~ 100%?

Answer: An overall probability of ~100% of a triggered landslide is improbable. Suppose all three faults have a 100% probability of producing their maximum-magnitude earthquake within 30 years:

$$P_{SA} = 1.0 \times 0.16 = 0.16$$

$$P_{MA} = 1.0 \times 0.20 = 0.20$$

$$P_{HR} = 1.0 \times 0.03 = 0.03$$

In this unlikely sequence we would have

$$P = 1 - [(1 - 0.16)(1 - 0.2)(1 - 0.03)] = 0.35 \quad \text{or} \quad 35\%$$

An overall probability of ~ 100% would, in addition, require much higher probabilities assigned to the likelihood of landslides being triggered.

Index

No Longer
the Property of
Bluffton University